Quality Beyond Six Sigma

Ron Basu and Nevan Wright

Routledge
Taylor & Francis Group
LONDON AND NEW YORK

First published by Butterworth-Heinemann

First published 2003

This edition published 2011 by Routledge
2 Park Square, Milton Park, Abingdon, Oxon OX14 4RN
711 Third Avenue, New York, NY 10017, USA

Routledge is an imprint of the Taylor & Francis Group, an informa business

British Library Cataloguing in Publication Data
A catalogue record for this book is available from the British Library

Library of Congress Cataloguing in Publication Data
A catalogue record for this book is available from the Library of Congress

ISBN - 978 0 7506 5561 3

Typeset by Replika Press Pvt. Ltd, India

Contents

Foreword

Since the early 1980s, in the 'Western World' we have been in what I have called a quality revolution. Based on the simple premise that organizations of all kinds exist mainly to serve the needs of the customers of their products or services, good quality management has assumed great importance. Competitive pressures on companies and Government demands on the public sector have driven the need to find more effective and efficient approaches to managing businesses and non-profit making organizations.

In the early days of the realization that improved quality was vital to the survival of many companies, especially in manufacturing, senior managers were made aware, through national campaigns and award programmes, that the basic elements had to be right. They learned through adoption of quality management systems, the involvement of improvement teams and the use of quality tools, that improved business **performance** could be achieved only through better **planning**, capable **processes** and the involvement of **people**. These are the basic elements of a **Total Quality Management** ((TQM) approach and this has not changed no matter how many sophisticated approaches and techniques come along.

The development of TQM has seen the introduction and adoption of many dialects and components, including quality circles, international systems and standards, statistical process control (SPC), business process re-engineering (BPR), lean manufacturing, continuous improvement, benchmarking and business excellence.

An approach finding favour in some companies was Six Sigma, most famously used in Motorola, General Electric and Allied Signal. This operationalized TQM into a project-based system, based on delivering tangible business benefits, often directly to the bottom line. Strange combinations of the various approaches have led to Lean Sigma and other company specific acronyms such as 'Statistically Based Continuous Improvement (SBCI)'.

The authors of this book have looked at the history of what I call TQM and developed another approach – Fit Sigma – which they hope will address some of the failures in the implementation of earlier projects and programmes, particularly in smaller companies and service organizations. In Fit Sigma the authors offer a holistic approach that fits the needs of all types of businesses and sustains improved performance. I wish them well with this book, but readers should recognize that the key element of any successful improvement management scheme is real and total commitment to the approach, alignment with the business strategies and dedicated follow through in the implementation.

John Oakland
Executive Chairman
Oakland Consulting plc (www.oaklandconsulting.com)
and European Centre *for* Business Excellence (www.ecforbe.com)
Professor of Business Excellence and Quality Management, Leeds University Business School

Professor Oakland is author of *Total Quality Management – text with cases, Statistical Process Control* and *Total Organizational Excellence.*

Preface

Whilst passing through Miami airport en route to Mexico City, Ron came across an article on Six Sigma in *USA Today*, 21 July 1998. It read: 'Today, depending on whom you listen to, Six Sigma is either a revolution slashing trillions of dollars from corporate inefficiency or it's the most maddening management fad yet devised to keep front-line workers too busy collecting data to do their jobs'. At that time Ron was coordinating a Global MRPII programme between all manufacturing sites of GlaxoWellcome, including the Xochimilco site in Mexico. The Global Manufacturing and Supply Division of GlaxoWellcome was considering a 'LeanSigma' initiative, which was meant to be a hybrid of Six Sigma and Lean Manufacturing. It struck Ron that the message in *USA Today* reflected not just the doubts (or expectations) in the minds of colleagues, but perhaps also those of quality practitioners worldwide.

These doubts or expectations addressed many questions. Isn't Six Sigma simply another fad, or just a repackaged form of TQM? It appears to be successful in large organizations like Motorola and General Electric, but can a small firm support such a programme? How can we apply Six Sigma methodology, originating from manufacturing operations, to the far larger market of the service sector? Like any good product, Six Sigma will have a finite lifecycle – so what is next? Surely one big question is, how can we sustain the benefits in the longer term? It is good to be 'lean' but isn't it better to be 'fit', to stay agile? The idea of writing *Quality Beyond Six Sigma* to address these issues was mentally conceived at Miami airport, and the concept of FIT SIGMA™* was born.

Ron nurtured the concept of FIT SIGMA for about two years, and then the opportunity came to write the book. In 2000 Nevan Wright, Ron's co-author for *Total Manufacturing Solutions*, returned to England from New Zealand to complete his PhD research at Henley Management College, and met up with Ron. Nevan has carried out extensive research into total quality and service performance, and is also the author of *The Management of Service Operations*. From our previous partnership we knew that we complement each other and share the same philosophy *re.* quality and continuous improvement, and thus we found a perfect fit for the FIT SIGMA project.

The central theme of this book is to provide a practical approach for FIT

---

*FIT SIGMA™ is a trademark of Performance Excellence Ltd, UK, copyright Ron Basu.

SIGMA, supported by case studies and phased action plans. The three distinctive features of FIT SIGMA are:

1. Fitness for purpose
2. Sigma (Σ) for improvement and integration
3. Fitness for sustainability.

FIT SIGMA has three important aspects. The first is that with FIT SIGMA we identify key areas where zero defects are essential plus areas where zero defects are possible, but we also recognize that there are areas where zero defects are not essential or practical. We believe that Sigma should fit the requirements of the organization, rather than the organization striving to fit an imposed mathematical formula. Using our approach Six Sigma can be made to fit any type or size of organization, whether in manufacturing or services. The second focuses on deploying a holistic approach to Six Sigma, with a conscious shift from variation (σ) to integration (Σ) across every function. The third aspect is sustaining the benefits gained. Many an organization has adopted Six Sigma, and the same applies to other quality initiatives, and found that the initial enthusiasm and successes gained were not sustained. In *Quality Beyond Six Sigma* we show how to sustain the benefits gained, and how to maintain the enthusiasm of the staff of the organization. We call this 'keeping fit'. Thus FIT SIGMA (or FIT Σ) means a quality system that first fits the needs of the organization (fitness for purpose), secondly is a holistic approach that integrates (Σ) all functions, and thirdly keeps the organization fit. Once an organization is fit, its culture will be developed to such an extent that staff will be striving for organizational *kaizen* – i.e. the organization as a whole will continuously become even fitter!

Senior executives and managers of all types and sizes of organizations and management consultants and students of all disciplines will find this book a stimulating guide to quality and operational excellence.

Lumen accipe et imperiti – Take the light and pass it on.

Ron Basu and Nevan Wright
Gerrards Cross, England and Auckland, New Zealand, June 2002

Acknowledgements

I wish to acknowledge the support and encouragement of my colleagues and students at Henley Management College in England, Europe, Africa and Asia, and of my colleagues and students at the Auckland University of Technology in New Zealand.

In particular, I wish to acknowledge the encouragement of Professor Ray Wild for my research at Henley Management College.

I again thank Joy, my wife, for her support and patience. I also wish to thank Natalie White for her assistance with research and with editing parts of the manuscript.

As always, working with Ron has been a pleasure.

Finally my thanks go to Maggie Smith, Nicki Kear, Deena Burgess and other staff of Butterworth-Heinemann.

Nevan Wright

Thomas A. Edison once said: 'Your idea has to be original in the adaptation to the problem you are currently working on'. By definition, *Quality Beyond Six Sigma* is a continuation of the quality movement and its originality is in its application to current business problems. Many of the tools and techniques in the book are those of other writers and quality gurus, and in that sense I am grateful for the work of our predecessors, some of whom are legends in the quality business. There is also another group of people to whom I am grateful for trying out these tools and ideas, and these are the practitioners and managers with whom I worked and learned in Unilever and GlaxoWellcome for nearly two decades and, more recently, the MBA students of Henley Management College.

I wish to acknowledge the friendship and invaluable input of my co-author Nevan Wright in the preparation of this book. I am grateful for the generous contributions to various case examples by my contacts through Six Sigma conferences – in particular, Kathleen Bader and Jeff Schatzer of the Dow Chemical Company and Rob Hardeman of Seagate Technology. The support and positive comments of Peter Race (Henley Management College) are greatly appreciated.

We are fortunate to have continued support from the team at Butterworth-Heinemann, especially from Maggie Smith, Nicki Kear and Deena Burgess.

Finally, this project could not be completed without the encouragement of my wife Moira and daughter Bonnie. Even my son Robi, in spite of the inherent scepticism of youth, presented me with *Jack* by Jack Welch to demonstrate his tacit support.

Ron Basu

1 Why FIT SIGMA™?

Men my brothers, men the workers, ever reaping something new:
That which they have done but earnest of the things that they shall do.

Tennyson

Introduction

This chapter considers the world of change and the need for organizations to be aware of the factors required to sustain competitive advantage from the introduction of change programmes.

A competitive world

This is a competitive world. The pace of change is increasing, and businesses are continuously being disrupted by external factors. In recent times the biggest external factor has been e-commerce or e-business. The spectacular rise in 1999 and fall in 2000 of so many dot.com companies showed that without substance no business will survive. When a large bubble bursts, innocent bystanders will feel some effect. E-business has actually been around for many years, and organizations with substance have benefited vastly from the intelligent use of information technology. The most successful pioneer in e-business is arguably the banking industry; likewise, the success of bar coding in the supply chain cannot be denied. In 2001 e-business made another spectacular advance, with the formation of large business-to-business alliances. Business to business (B to B) took on a new meaning with the advent of the Covisint alliance between Daimler/Chrysler, Ford and General Motors. It is reported that the alliance of these three major (and fiercely competitive) organizations has a purchasing power of $300 billion per annum. The interesting phenomenon is that such fiercely competitive organizations have been able to form an alliance at all! Other industry groups have been quick to follow – for example, the oil companies, the aeronautical industry, the computer industry (led by IBM), and alliances of fast-moving consumer goods distributors.

Change is here to stay; it comes quickly and from unexpected quarters. The challenge for all businesses is to find the benefits of change, with the aims of generating real revenue and delivering enhanced value to customers. The B-to-B alliances are at early stages of development and the benefits are yet to be realized, but the advantages of bar coding, supply chain management and electronic banking are obvious and are now taken for granted. The failed dot.com companies clearly did not produce real revenue and generally did not provide the benefits promised to customers, and thus their early apparent success was not sustainable.

There are similar stories of unsustainable improvements in traditional businesses in the 'old economy' (prior to e-business). In spite of the demonstrated benefits of many improvement techniques, such as Total Quality Management, Business Process Re-engineering and Six Sigma, many attempts to implement and sustain improvements have fizzled out, not with a bang but with a whimper. What is more puzzling is that some companies who successfully implemented a quality initiative and initially reported substantial improvements have subsequently experienced overall drops in performance and profit, resulting in lay offs and lowered employee morale. For example, Motorola, the originator of Six Sigma, announced in 1998 that its second quarter profit was almost non-existent and that it was reducing its staff of 150 000 by 15 000. At the time of writing (May 2002) the situation for Motorola has not improved. The actual number of job cuts since August 2000 is 48 400 (almost one-third of the work force); for the year ended 2001 the company reported its first operating loss in 71 years; and the stock value has declined by 73 per cent over the last two years.

Why successes are not sustained

There are many hidden reasons why organizations (in both the old and the new economies) do not sustain the initial successes gained from improvement initiatives.

One main factor is the lack of solid measures. All companies have one key measure; the return on assets, or the 'bottom line'. However, the bottom line is a historical measurement – no matter how good the accounting system, by the time the bottom line is known it is too late to influence the result. The bottom line is in itself a measure of the result, and for many, such as bankers, investors and the share market, it is the result. Thus most organizations considering a new management technique or quality initiative, such as balanced score cards, business process re-engineering, benchmarking, just-in-time systems or what ever else is the flavour of the month, are in the main looking to save costs so as to improve the bottom line. Lip service is given to customer service (as espoused in the mission statement), but the reality is to get the costs down and the bottom line up. The measures must be truly balanced and underpinned by a formal process of periodic assessment and senior management review.

The second main factor is the apparently finite lifecycle of change programme 'products' such as TQM and Six Sigma. Turner (1999) finds that a typical scenario for the implementation and maintenance of a quality programme has a lifecycle of approximately two and a half years. He believes that initially enthusiasm is high and staff are very committed to a new way of working, but as time progresses setbacks may occur, unanticipated problems may arise, or perhaps the novelty simply wears off. Another key factor is the lack of a holistic approach to the management of organizations where economic, social, and environmental criteria of the business are valued for its sustainability. This fundamental strategy has been described in detail in Total Manufacturing Solutions (Basu and Wright, 1998). The Department of Trade and Industry of the British Government has sponsored a project called SIGMA – Sustainability: Guidelines for Management – under the direction of the British Standards Institute to promote the holistic principles of sustainability. The aim of the SIGMA project is given as 'to build the capacity of organizations to meet their business and other institutional objectives by more effectively addressing their social, environmental and economic dilemmas, threats and opportunities' (see www.ProjestSIGMA.com, December 2001).

Six Sigma

Six Sigma is an approach that takes a whole system approach to improvement of quality and customer service so as to improve the bottom line. The Six Sigma concept matured between 1985 and 1986, and grew out of various quality initiatives at Motorola. Like most quality initiatives since the days of Dr W. Edwards Deming in the 1960s, and in particular the concept of Total Quality Management (TQM), Six Sigma requires a total culture throughout an organization whereby everyone at all levels has a passion for continuous improvement with the ultimate aim of achieving virtual perfection. The difference with Six Sigma is the setting of a performance level that equates to 3.4 defects per 1 million opportunities. To know if Six Sigma has been achieved a common language is needed throughout the organization (at all levels and within each function), and common uniform measurement techniques of quality are necessary. The overall Six Sigma philosophy has a goal of total customer satisfaction.

In 2000, Ron Basu surveyed the following leading companies who had adopted the Six Sigma approach to quality:

- Motorola
- Allied Signal (Honeywell)
- General Electric
- Raytheon
- DuPont Teijn
- Bombadier Shorts
- Seagate Technology

- Foxboro (Invensys)
- Norando
- Ericson
- Dow Chemical.

Ron found that the main driver leading to the application of Six Sigma to a company is cost saving rather than customer satisfaction! In coming to this conclusion Ron benefited from informal networking with members of the above companies and also leading consulting groups such as Air Academy Associates, Rath and Strong, Price Waterhouse Cooper, Iomega, and Cambridge Management Consulting.

The surveyed companies reported between them a long list of intangible and indirect benefits. However, these benefits did not seem to be supported by any employee or customer surveys (Basu, unpublished paper).

Nonetheless, very real results from the adoption of Six Sigma continue to be reported. For example, in 1997 Citibank undertook a Six Sigma initiative and after just three years it was reported that defects had reduced by ten times (see Erwin and Douglas, 2000, for details). Likewise, General Electric reported that $300 million invested in 1997 in Six Sigma delivered between $400 million and $500 million savings, with additional incremental margins of $100 to $200 million. Wipro Corporation in India says that two years after starting in 1999, defects were reduced to such an extent as to realize a gain of eight times over the investment in Six Sigma.

Against this background let us examine the evolution of the total quality improvement process (also known as operational excellence) from ad hoc improvement, to TQM, to Six Sigma, up to Lean Sigma. Building on the success factors of these processes the key question is, how do we sustain the acquired benefits? The answer lies with FIT SIGMA.

Lean Sigma

Basically, if accuracy in the order of 3.4 defects per million opportunities is added to the key ingredient of quality, and this is implemented across the business with an intensive education and training programme, we have Six Sigma. We will now look at lean enterprise, which is in fact an updated version of industrial engineering. With lean enterprise the focus is on delivered value as seen by the customer. The aim is to eliminate all non-value-adding activities (wasted effort, wasted materials) for each product and process along the value chain. The value chain begins with the supplier and the supplier's supplier, and flows through the transformation process to the organization's direct customer, and finally to the customer's customer. The value chain relies on two-way communication from the end user back to the original supplier. The integration of Six Sigma and the lean enterprise approach gives Lean Sigma.

Incremental is not enough

What frightens people is the target of Six Sigma – 3.4 defects per million opportunities is almost perfection, and seems impossible or even unnecessary. Instead, some management hide behind the concept of continuous improvement. However, almost all organizations today are striving to make continuous gradual or incremental improvements, and these companies obviously include your competitors! Incremental today is not enough; it is too slow, and only keeps pace with mediocrity. What do you do if your main competitor announces that it has reduced expenses by 10 per cent, it will deliver a markedly improved product in half the time and increase the level of service, and will not increase the price? Erwin and Douglas (2000) cite Craig Erwin of Motorola:

> *Before Six Sigma, we were interested in continuous improvement, but we tended to accept quality that mirrored our competitors. We were internally focused and accepted the argument that things couldn't be made better. When we started, many people thought Six Sigma was unrealistic.*

Ron Randall of Texas Instruments, in comparing his division (DSEG, now called Raytheon TI Systems) with Motorola, said:

> *... in addition to being impressed with the quantitative methods the moment that helped cement everyone's commitment was when DSEG looked at its products and compared them to similar ones from Motorola. We were less than Four Sigma AND Motorola was close to Six. We couldn't believe someone was 2000 times better than us. It really got our attention. We were always pursuing quality, we thought, but it was incremental.*

Motorola initially concentrated on applying Six Sigma to its manufacturing units. Bob Galvin, former chief executive of Motorola, now says that the lack of initial Six Sigma initiative in the non-manufacturing areas of Motorola was a mistake that cost Motorola at least $5 billion over a period of four years!

The new wave: quality beyond Sigma

FIT SIGMA is the new wave of Sigma. Lean Sigma provides agility and efficiency; FIT SIGMA also ensures sustainability. We call this maintaining fitness. FIT SIGMA also considers what is really required for a specific organization or operation. We will show that is not necessary for every operation to achieve the virtual perfection level of 3.4 errors per million opportunities – FIT SIGMA is what is fit for the operation. Not all organizations need the intensive and expensive 'all or nothing' investment required by the Six Sigma deployment plan.

Our philosophy is the adaptation of the Six Sigma approach to fit an organization's needs so as to maintain performance and organizational fitness.

The evolution of TQM to Six Sigma to Lean Sigma to FIT SIGMA

The evolution of TQM to FIT SIGMA is shown in Figure 1.1.

It began with Total Quality Management (TQM), as originated by Dr W. Edwards Deming. Dr Deming, an American statistician had great input into turning around Japanese industry after the Second World War. His efforts were so appreciated that the Japanese have an annual highly recognized quality award known as the Deming prize.

After his success in Japan, Dr Deming was not really recognized in the United States until the 1980s when he, in his eighties, was asked by the CEO of Ford to advise on how to get quality back into the car manufacturing industry. At that time American industry, in particular the automobile sector, was reeling from the influx of high quality and comparatively cheap Japanese products. Deming is credited for turning Ford around by introducing quality methods based on rigorous discipline in the factory, statistical methods, and a change in culture.

The change in culture required:

- Management to recognize that 90 per cent of all quality problems (faults, scrap and reworks) were the result of poor management and processes
- Workers and management to learn to trust each other so that everyone accepted that each had a personal responsibility for quality and improvement.

Following his successes in Detroit, for a time Dr Deming became the most widely sought-after management guru in America. Deming and the quality movement is discussed in some detail in Chapter 2.

Six Sigma began with Motorola under the leadership of Bob Galvin in the mid-1980s, and was not an entirely new technique. Six Sigma takes a handful of proven techniques from TQM and uses them to train a small group of in-house technical people to become Sigma 'Black Belts'. This training includes the use of advanced computer programs, which in themselves are not difficult to learn or apply. Advanced Black Belts become Master Black Belts, and Master Black Belts provide technical leadership for the Six Sigma program. Whereas Black Belts apply the mathematical statistical formulas, Master Black Belts must also understand the theory on which the statistical methods are based. Master Black Belts train Black Belts and Green Belts. Black Belts typically receive 160 hours of classroom instruction and one-on-one project coaching from Master Black Belts or from consultants; Green Belts are Six Sigma project leaders capable of forming and leading Six Sigma teams. Green

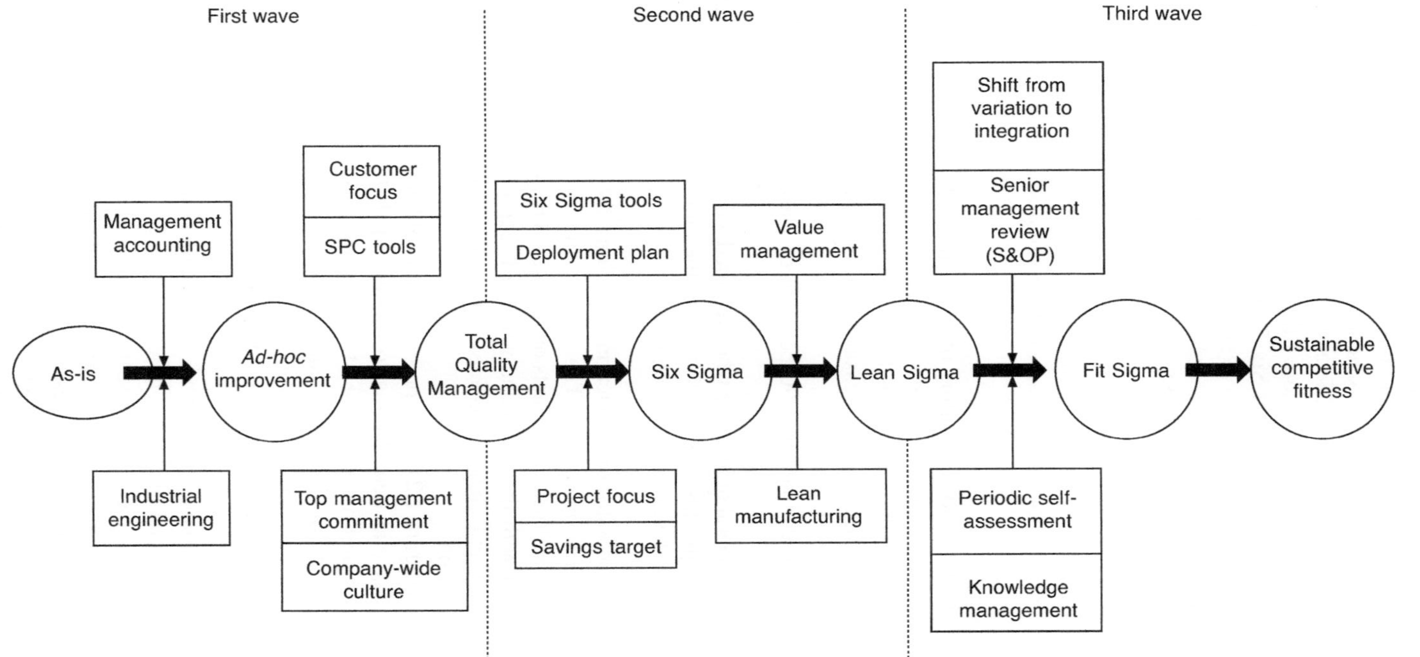

Figure 1.1 FIT SIGMA™ road map – the evolution of TQM to Fit Sigma.

Belt training usually consists of five days of classroom training, covering project management, quality control tools, problem solving and data analysis. Drawing on Deming's 'Plan = Do = Check = Act cycle (PDCA)', Six Sigma's performance model is: Define = Measure = Analyse = Improve = Control (DMAIC). DMAIC is explained more fully in Chapter 3.

It is important to note here that Six Sigma is a disciplined methodology and that it requires an infrastructure to assure that performance improvement initiatives are supported with the necessary resources.

The next wave in the FIT SIGMA evolution is Lean Sigma. Lean Sigma incorporates the lean production methods of the Japanese just-in-time approach synonymous with Toyota and made famous by Womack, Jones and Roos in their bestseller *The Machine That Changed The World* (Womack et al., 1990) Lean production aims for elimination of the seven *mudas* (non-value-adding activities):

1. Excess production (no stockpiling of finished goods)
2. Waiting (no buffer stocks between processes, no idle time)
3. Conveyance (movement is reduced to a minimum)
4. Motion (elimination of unnecessary motion, adoption of ergonomic principles)
5. Process (Deming claimed that 90 per cent of waste is due to inefficient processes)
6. Inventory (materials should arrive when required, go straight into production, and flow like water through the system to the end user)
7. Defects (the aim being zero defects).

Lean Sigma relates not just to production operations; the principles are equally applicable to service operations. The overall aim is to reduce waste and improve the delivery times of products or services. The predictable Six Sigma process combined with the speed and agility of lean provides solutions that give better, faster and cheaper business processes coupled with improved customer satisfaction.

FIT SIGMA (Figure 1.2) is the process that enables the dramatic bottom-line results of Six and Lean Sigma to be sustained. It ensures that where extensive training and development of skilled Sigma practitioners (Master, Black Belt and Green Belt) has been carried out, this is are not wasted and the benefits are secured for the long term.

FIT SIGMA adds the following features to Six and Lean Sigma:

- A formal senior management review at regular intervals, similar to the sales and operational planning process
- Periodic self-assessment with a structured checklist, which is recognized by a certificate or award, similar to the European Foundation of Quality Management or Baldridge process
- A continuous learning and management programme
- A whole systems approach across the entire organization.

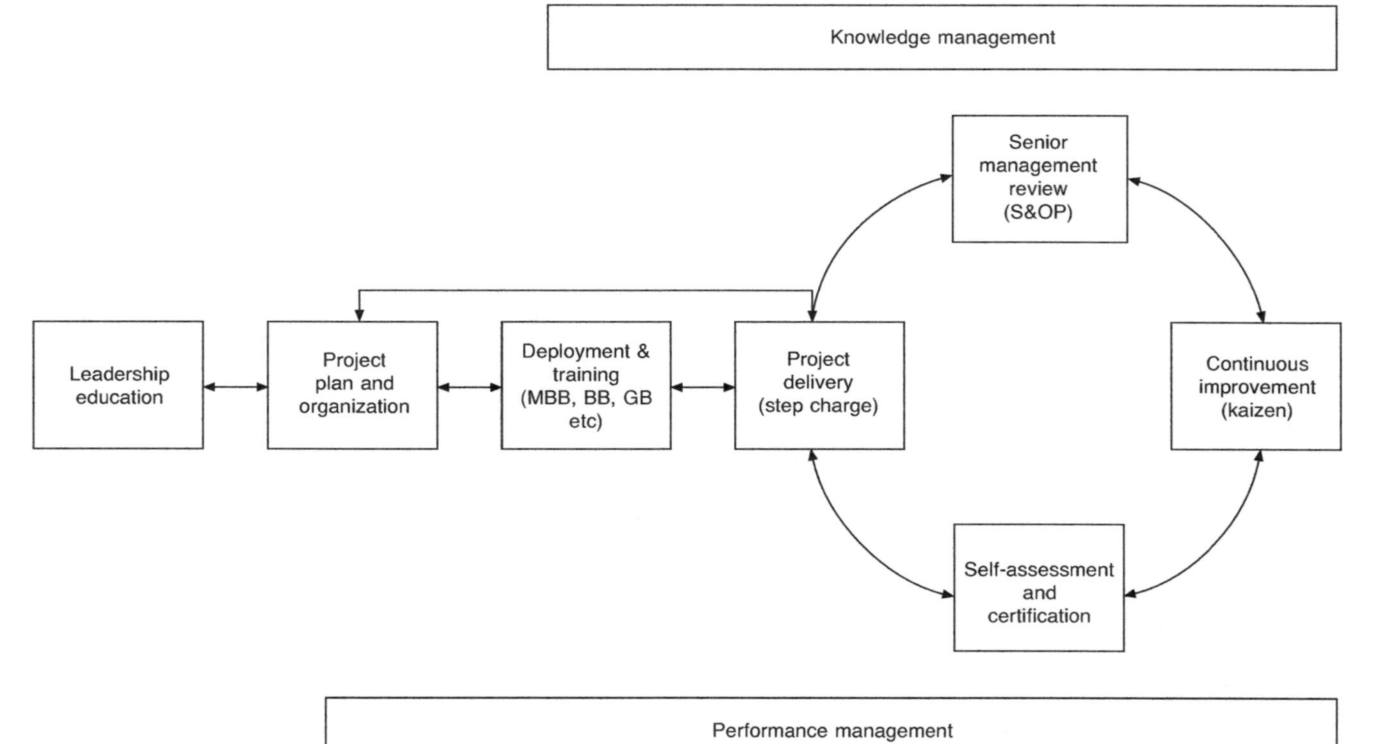

Figure 1.2 The Fit Sigma™ model.

Summary

FIT SIGMA is not a statistical tool; it is both a management philosophy and an improvement tool. The underlying philosophy is that of a total business-focused approach underpinned by continuous reviews and a knowledge-based culture to sustain a high level of performance. In order to implement the FIT SIGMA philosophy, a systematic process is necessary. The process is not a set of new or unknown tools. The tools are drawn from those that have been tried and proven in the successive waves of quality over the last 40 years, beginning with TQM, going on to Six Sigma and including Lean Sigma. The difference in FIT SIGMA is recognition of the need to sustain and retain successes. It is not a rigid programme in search of problems; it is an adaptable solution that can be tailored to fit any organization.

FIT SIGMA is not a magic formula; it is a total business philosophy, process and culture.

2

History of the quality movement

The bitterness of low quality is not forgotten
Nor can it be sweetened with low price.

Marquis De Lavant (1734)

Introduction

The Malcom Baldridge Award, the Deming Quality Award and The European Foundation of Quality have all served to give TQM a high profile. One count suggests that there are over 400 TQM tools and techniques (Pyzdek, 2000). This high profile has, however, paradoxically contributed to a level of scepticism, especially by middle managers and staff. Promises have not been realized, high-profile organizations that claimed to be practising TQM have gone into decline, and staff have seen slogans and mission statements published that focus on customer service and people coupled with TQM, followed by redundancies and drastic cuts in training budgets. This chapter discusses what is meant by quality and gives a historical overview of the development of quality thinking, beginning with Total Quality Management. It concludes with a summary of how FIT SIGMA™ builds on prior quality initiatives.

World class

The term world class is generally attributed to Hayes and Wheelwright (1984), who related best practice to German and Japanese firms competing in export markets. Schonberger (1986) used the term 'best practice' to describe manufacturers making rapid and continuous improvement. World class in the nineties was extended to include lean production (see Womack et al., 1990), referred to in Chapter 1.

Fry and co-workers (1994) and Harrison (1998) say best practice refers to any organization that performs as well as or better than the competition in quality, timeliness, flexibility and innovation. Knuckey and co-workers (1999, p. 23) explain that:

... the logic behind best practice is simple: because operational outcomes are a key contributor to competitiveness and business performance, and because best practice should improve operational outcomes, by implication good practice should lead to increased competitiveness. Best practice should lead to world class service.

Knuckey et al. (1999, p. 137), on behalf of the New Zealand Ministry of Commerce, found from research of 1173 New Zealand manufacturing firms that the 'main sources of competitive advantage' and 'best practice' is;

- goodwill and trust with suppliers and distributors
- trust, goodwill and commitment from employees to the firm's goals, and reputation with clients.

Why best practice and world class is essential

There is no doubt that people today are more travelled, better educated and consequently more discerning than ever before. Customers know what is on offer elsewhere, they have experienced it and their expectations have been raised by advertising and marketing. Likewise, shareholders and other financial stakeholders can be excused for wondering why the rapid technological advances of the last decade have not resulted in increased performance and higher returns on investment. At the same time, the well-publicized and promised benefits of technology have led customers to expect – even demand – improved products and service at less cost. Quality service, reliable products, value for money and accountability are now taken for granted. Competitors are global, standards are world class, and organizations that fail to meet world-class performance will soon be found out.

The Japanese approach to Total Quality Management

W.E. Deming

Total Quality Management (TQM) has its origins in Japan. In the 1960s, Japan went through a quality revolution. Prior to this, 'Made in Japan' meant cheap or shoddy consumer goods. The approach used in Japan in the 1950s and 1960s to improve quality standards was to employ consultants from America and Europe, and the most famous of these consultants was Dr W. Edwards Deming. Deming's philosophy was to establish the best current practices within an organization, to establish the best practice as standard procedure, and to train the workers in the best way. In this manner, everyone would be using the same best way. His approach was to involve everyone in the organization and win them over – he believed that quality was everyone's business. Deming said finding the best way meant collecting the facts, amassing data, setting standard procedures, measuring results, and getting prompt and accurate feedback on these results so as to eliminate variations to the standard. He saw this as a continuous cycle. Deming emphasized that people can only be won over if there is trust at all levels. This means that management must be prepared to allow and encourage employees to take responsibility, and employees must be prepared to accept responsibility. Employee participation, through understanding objectives, processes and contributing through improvement suggestions, is a serious part of the Deming philosophy. He claimed that cultivating the know-how of employees was 98 per cent of the quality challenge – as Gabor (2000) says, Deming has been criticized for hyperbole! However, Gabor adds (p. 293), quoting a Ford engineer, 'Deming understood that you can't turn quality on like a spigot {tap}. It's a culture, a lifestyle within a company'. The first of Deming's fourteen points of quality is 'Create constancy of purpose toward improvement of product and service', and his second point is 'Adopt the new philosophy . . . management must take leadership for change' (Deming, 1986; Walton, 1986; Gabor, 2000). The overall philosophy of TQM is one of incremental and continuous improvement, not revolution.

Deming's fourteen points of quality

No section on Dr Deming is complete without reference to his famous fourteen points of quality (the comments in parentheses are our notes, and not direct quotations of Deming):

1. Create consistency of purpose toward improvement of product and service.
2. Adopt the new philosophy (management has to learn its responsibilities and to take leadership. It is difficult for management to accept that 90 per cent of problems lie with management and the process).
3. Cease dependence on inspection to achieve quality (supervision and supervisors' wages do not add value, they are an extra cost; far better if staff take responsibility and supervise themselves. Deming also added that if quality is built into the design or process, then inspection will not be necessary).
4. End the practice of awarding business on the basis of the price tag (the cheaper the price, the higher the number of failures. Move to dedicated suppliers, and value reliability, delivery on time and quality).
5. Improve constantly and forever the system of production and service (this is an extension of the Japanese philosophy of kaizen, whereby not a day should go by without some incremental improvement within the organization).
6. Institute training on the job (become a learning organization with a willingness to share knowledge).
7. Institute leadership (everyone at all levels, especially supervisors, should be team leaders and not disciplinarians. Everyone should be encouraged

to develop self-leadership. Quality is too important to be left to management).

8. Drive out fear (encourage people to admit mistakes; the aim is to fix not to punish. However it is expected that people won't go on making the same mistakes!)
9. Break down barriers between departments (eliminate suspicion between departments. There needs to be clear objectives, with everyone striving to work for the common good).
10. Eliminate slogans, exhortations and targets for the workforce (there is no use asking for zero defects if the process or the product design is not perfect; 10 per cent across-the-board cost reduction demands are poor for morale if they are not possible).
11. Eliminate work standards – quotas – on the factory floor (e.g. 100 pieces per hour with a bonus for a 110 will result in 110 pieces, but not necessarily in quality products. The focus will be on output numbers rather than quality. If the worker is encouraged to consider quality, 95 high-quality pieces per hour will be worth more than 110 if 15 (of the 110) are subsequently rejected or returned by the customer).
12. (a) Remove barriers that rob the worker of the right to pride of workmanship (give them the right tools, right materials, right processes and comfortable working conditions; treat them with respect).
    (b) Remove barriers that rob people in management and in engineering of their right to pride in craftsmanship (this includes appraisal systems that reward on bottom-line results and keeping expense budgets down, and ignore customer satisfaction. If cost is the only driver, then training, maintenance and customer service etc. will suffer).
13. Institute a programme of education and self-improvement (encourage staff to seek higher educational qualifications; become a knowledge-based organization).
14. Put everybody in the company to work to accomplish the transformation (change of culture is difficult to achieve. Dr Deming saw that everyone has to be involved in transforming the culture of an organization).

Dr Joseph M. Juran

Deming was not the only guru of quality used by the Japanese. Dr Joseph M. Juran was also associated with Japan's emergence as the benchmark for quality of products. Juran was, like Deming, an American statistician, and there are similarities between his work and that of Deming. Above all, both men highlight managerial responsibility for quality. Arguably Juran was the first guru to emphasize that quality is achieved by communication. The Juran trilogy for quality is planning, control and improvement (Juran, 1989). His approach includes an annual plan for quality improvement and cost reduction, and continuous education on quality. Juran's foundations are still valid, and are embedded within Six Sigma and Lean Sigma and our FIT SIGMA philosophies. Juran uses the term 'quality control', but this does not refer to the post-production inspection that passes for quality control in many organizations. He argues, and few would disagree, that inspection at the end of the line, post-production, is too late to prevent errors. Juran says that quality monitoring needs to be performed during the production process to ensure that mistakes do not occur and that the system is operating effectively. He does this by examining the relationship between the process variables and the resultant product. Once these relationships have been determined by statistical experiment, the process variables can be monitored using statistical methods. Juran adds that the role of the upper management is more than making policies; they have to show leadership through action – they have to walk the talk, not just give orders and set targets. He says that quality is not free and that investment (often substantial) in training, including statistical analysis, is needed at all levels of the organization. Juran also believed in the use of quality circles. As he describes them, quality circles are small teams of staff with a common interest who are brought together to solve quality problems. Our constituents for a successful quality circle are discussed later in this chapter.

As can be seen from this brief synopsis of Dr Juran's philosophies, there is nothing that he says that is not complementary with Six Sigma.

Armand V. Feigenbaum

Feigenbaum is recognized for his work in raising quality awareness in the USA. He was General Electric's worldwide chief of manufacturing operations for a decade until the late 1960s. The term Total Quality Management originated from his book Total Quality Control, first published in 1961 (Feigenbaum, 1983). Feigenbaum states that Total Quality Control has an organization-wide impact, which involves managerial and technical implementation of customer-orientated quality activities as a prime responsibility of general management and of the main-line operations of marketing, engineering, production, industrial relations, finance and service as well as of the quality-control function itself. He adds that a quality system is the agreed company-wide operating work structure, documented in integrated technical and managerial procedures, for guiding the coordinated actions of the people, the machines, and company-wide communication in the most practical ways, with the focus on customer quality satisfaction.

Feigenbaum was one of the first writers to recognize that quality must be determined from the customer's perspective, and NOT the designer's (or the engineer's or the marketing department's) concept of what quality is.

Feigenbaum also said that the best does not mean outright excellence, but means the best for satisfying certain customer conditions. In other words, as in FIT SIGMA, 'best' means sufficiently good to meet the circumstances. Feigenbaum, like Deming and Juran, found that measurement is necessary, but whereas Deming and Juran tended to measure production and outputs Feigenbaum concentrated on measurement to evaluate whether good service and product met the desired level of customer satisfaction.

Dale (1999) believed that Feigenbaum's major contribution to quality was to recognize that the three major categories of cost are appraisal, prevention, and cost of failure. According to Feigenbaum, the goal of quality improvement is to reduce the total cost of quality from the often quoted 25–30 per cent of cost of sales (a huge percentage when you think about it) to as low a percentage as possible. Thus Feigenbaum takes a very financial approach to the cost of quality.

To summarize, Feigenbaum's approach is not substantially different to that of Deming and Juran but his emphasis is different – he defines quality in product and service from the customer's perspective, he does not aim for the outright best, and he takes a financial approach to the cost of quality.

P.B. Crosby

P.B. Crosby, a guru of the late 1970s, was the populist who 'sold' the concept of total quality management and 'zero defects' to the USA. Although the zero defects concept sounds very much like Six Sigma, in fact Crosby takes a very much softer approach than does Deming, Juran, Feigenbaum or Six Sigma. His concept of zero defects is based on the assumption that it is always cheaper to do things right the first time, and quality is conformance to requirements. Note the wording 'conformance to requirements' – thus any product that conforms to requirements, even where requirements are specified at less than perfection, is deemed to be defect free.

Crosby developed the concept of non-conformance when recording the cost of quality. Non-conformance includes the costs of waste and scrap, down time due to poor maintenance, putting things right, product recall, replacement and, at worst, legal advice. All these can be measured, and, according to Crosby, cost of non-conformance 'can be as much as 20 per cent of manufacturing sales and 30 per cent of operating costs in service industries'. To this list we add the costs that can't be measured when things go wrong, such as market reputation and consequential lost sales, and lost management time wasted on troubleshooting and customer relations – time and energy that could have been well spent on planning and strategy etc.

Crosby is famous for saying that quality is free (Crosby, 1979). He emphasized cultural and behavioural issues ahead of the statistical approach of Deming and Feigenbaum. Crosby was saying that if staff have the right attitude, know what the standards are and do things right first time every time, the cost of conformance is free. The flow-on effect is that motivated workers go further than just doing things right; they detect problems in advance, are proactive in correcting situations, and are quick to suggest improvements. Crosby concluded that workers should not be blamed for errors, but rather that management should take the lead and the workers will follow. Crosby suggests that 85 per cent of quality problems are within management control (Deming put this figure at 90 per cent).

What of the Japanese?

The more important of the Japanese writers on quality are Genichi Taguchi, Ishikawa, Shingo and Imai. Also, Toyota is of course widely cited as the epitome of lean production (Womack et al., 1990).

Of the Japanese approaches to quality, the Taguchi methods have been the most widely adopted in America and Europe. Taguchi, an electrical engineer, used an experimental technique to assess the impact of many parameters on a single output. His method was developed during his work rebuilding the Japanese telephone system in the 1970s. His approach to quality control is focused on 'off line' or loss of function (derived from telephone system failures).

The Taguchi approach is to:

- Determine the existing quality level measured in the incidence of down time, which he called 'off line'
- Improve the quality level by parameter and tolerance design
- Monitor the quality level by using statistical process control to show upper and lower level variances.

Taguchi advocates three stages of quality design, namely:

1. System design – this is the development of the basic system, which involves experimentation with materials and the testing of feasibility with prototypes. Obviously, technical/scientific knowledge is a requisite.
2. Parameter design – this begins with establishing the optimum levels for control factors so that the product or process is least sensitive to the effect of changes of conditions (i.e. the system is robust). This stage includes experimentation, with the emphasis on using low-cost materials and processes.
3. Tolerance design – this includes setting numerical values (factors) for upper service levels and lower acceptable service levels, and reconciling the choice of factors in product design. In turn, this includes comparison of costs by experimenting with low-cost materials and consideration of more expensive materials to reduce the tolerance gap. Design includes process design and product design; process design includes choosing the upper and lower parameters of service, and product design includes reconciling the choice of materials against the desired service level parameters.

Taguchi promotes three stages in developing quality in the design of product or systems:

1. Determine the quality level, as expressed in his loss function concept
2. Improve the quality level in a cost-effective manner by parameter and tolerance design

3. Monitor the quality of performance by use of feedback and statistical control.

As Ferguson and Dale (1999, p. 350) say:

> *Taguchi has raised the awareness of engineers and technical staff to the fact that many of the problems associated with design, production costs and process control can be resolved using experimental design and analysis methods. This contribution to both awareness and the knowledge base of the subject should not be overlooked.*

Basu and Wright, and Total Quality Management

Basu and Wright (1998) identify a hierarchy of quality management that has four levels: inspection, control, assurance and Total Quality Management (TQM).

Quality inspection and quality control rely on supervision to make sure that no mistakes are made. The most basic approach to quality is inspection, detection and correction of errors. The next level, quality control, is to inspect, correct, investigate and find the causes of problems and take actions to prevent errors re-occurring. Both methods rely on supervision and inspection. The third level, quality assurance, includes the setting of standards with documentation and also the documentation of the method of checking against the specified standards. Quality assurance generally also includes third-party approval from a recognized authority, such as found with the ISO 9000 series. With quality assurance, inspection and control are still the basic approach, but in addition a comprehensive quality manual, the recording of quality costs, and perhaps the use of statistical process control and sampling techniques for random and the overall auditing of quality systems would be expected.

Quality inspection and control and quality assurance are aimed at achieving an agreed consistent level of quality, first by testing and inspection, then by rigid conformance to standards and procedures, and finally by efforts to eliminate causes of errors so that the defined accepted level is achieved. We see this as a cold and sterile approach to quality. It implies that once a sufficient level of quality has been achieved then, apart from maintaining that level (which in itself might to increase productivity. Supervisors were employed to maintain the best method. Workers were not expected to make suggestions; their job was to do what they were told while management did the thinking.) It implies that the bosses know what is best; they set the standards, and they inspect and control to see that the standards are adhered to. This does not mean that management is not taking into account what the customer wants or is ignoring what the competition is doing, it just means that the managers believe they know what is best and know how this can be achieved.

Total Quality Management (TQM) is on a different plane. Total Quality Management does, of course, include all the previous levels of setting standards and the means of measuring conformance to standards. In doing this, Statistical

Process Control (SPC) may be used; systems will be documented, and accurate and timely feedback of results will be given. With TQM ISO accreditation might be sought, but an organization that has truly embraced TQM does not need the ISO stamp of approval. ISO is discussed briefly later in this chapter.

Any organization aspiring to TQM will have a vision of quality that goes far beyond mere conformance to a standard. TQM requires a culture whereby every member of the organization believes that not one day should go by without the organization in some way improving the quality of its goods and services. The vision of TQM must begin with the chief executive. If the chief executive does not have a passion for quality and continuous improvement, and if this passion cannot be transmitted down through the organization, then, paradoxically, the ongoing driving force will be from the bottom up.

Figure 2.1 depicts a TQM culture wherein management has the vision, which is communicated to and accepted by all levels of the organization. Once the quality culture has been ingrained in the organization, the ongoing driving force is 'bottom up'.

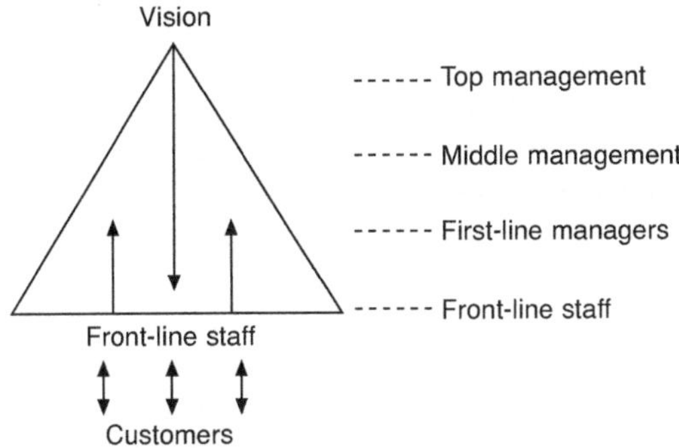

Figure 2.1 Quality and the driving force.

Generally it is the lower-paid members (shop assistants, sales representatives, telephone operators, van drivers, accounts clerks) of the organization who will be physically interfacing with the customers or providing the service, and it is their attitude and level of helpfulness that will determine the overall perception of quality by the customer. These workers have a huge part to play in how the customer perceives an organization. It is on the lower levels, then, that an organization must rely for the continuing daily level of quality.

Once the culture of quality has become ingrained, quality will be driven from the bottom up – whether by the factory worker or the sales assistant – rather than achieved by direction or control from the top.

Management will naturally continue to be responsible for planning and for providing the resources to enable the workers to do the job. However, unless the machine operator, shop assistant, telephone operator, cleaner, van driver and junior account clerk are fully committed to quality, TQM will never happen.

TQM also goes beyond the staff of the organization; it involves suppliers, customers and the general public.

American and European approaches to Total Quality Management

We have discussed the Japanese overall approach to TQM in the section relating to 'quality' gurus. As we have seen, the approaches that emanated from one culture, such as the Taguchi method, have crossed national boundaries. Likewise, the principles of TQM as practised in Japan were picked up by Feigenbaum and Crosby in the USA, and rapidly spread through the subcontinent, Europe and the rest of the world.

In the USA and in Europe, senior management, bankers and investors have a morbid fascination with share prices. They consequently feel pressure to meet short-term six-monthly targets of interim and annual reports, which are widely publicized and scrutinized. On each occasion that a report or statement is made, they must show a healthy bottom line or at least provide a promise of better results in the short-term forseeable future. Thus with the share price hanging like the sword of Damocles over their heads, the tendency is to look for instant results and quick fixes. The timeframe is short term, and if results are not readily apparent there will be a move to some other solution. The Japanese, however, know that success is rarely an overnight phenomenon.

The implementation of TQM, because it requires a total change in management thinking and a major change in culture, will take years to internalize. Thus with some organizations, because results are not instant, TQM has lost favour. Even where some positive results become apparent in a short space of time, they may not always seem to be major. How, though, can you tell if there have been benefits, and if they are significant or not?

If after adopting TQM an organization is still in business and the results are slightly up on the previous year, is this something to be excited about? Maybe the shareholders won't see this as a triumph, but it may well be. If the organization had not begun its quality revolution, perhaps the results would have been much worse.

John Oakland

In the UK, Professor Oakland is recognized as the leading light of Total Quality Management. His particular brand of TQM is essentially pragmatic, and includes a whole systems approach without relying on either quantitative or qualitative measures. It has been suggested that he leans towards qualitative aspects, i.e. the issues of culture, communication and teamwork. Some might refer to these as the 'softer issues', as it is difficult to quantify in 'hard' statistical terms a level of culture or teamwork. Like many writers, Oakland stresses the importance of these issues and offers a culture change cycle based on psychometrics such as MBTI and FIROB (Oakland, 2000).

Oakland's ten points of senior management

1. Make a long-term commitment
2. Change the culture to 'right first time'
3. Train the people to understand the customer–supplier relationship
4. Buy products and service on total cost (like Deming, Oakland is saying that the purchase price is not the final cost; total cost includes performance, running costs and repairs and maintenance costs)
5. Recognize that systems improvements must be managed
6. Adopt modern methods of supervision and training, and eliminate fear (the supervisor has to move from a strict disciplinarian role somewhat towards a mentoring role – guiding and supporting, not kicking butt and taking names)
7. Eliminate barriers, manage processes, improve communication and teamwork (encourage cross-functional department support, not defensive silo mentalities)
8. Eliminate arbitrary goals and standards based only on numbers, eliminate barriers to pride of work, use correct methods to get the facts and do not accept fiction or hearsay.
9. Constantly educate and train and use in-house experts where possible (bearing in mind that Oakland himself heads a consulting group)
10. Utilize a systematic approach to TQM implementation.

Referring back to Deming's fourteen points, it can be seen that Oakland's ten points reinforce rather than significantly add to TQM. Oakland has, however, applied a set of implementation tools known as Quality Function Deployment (QFD) to create a 'Goal Deployment' approach to aligning TQM with the business strategy.

QFD began in Japan in 1966, and Dr Yoji Akao is the recognized guru. QFD is a systematic approach to the design of a product or service so that customers' needs are met first time and every time. The approach includes forming teams of staff from across the functions of an organization to find out customers' needs and arrange how to meet them. QFD when applied to product/service design is achieved by:

- Market research
- Basic research
- Invention
- Concept design
- Prototype testing
- Final product or service testing
- After-sales service.

Oakland's QFD-based Goal Deployment approach requires top management commitment, and teamwork across the organization, and good process management. Oakland has been credited in the UK with successfully introducing

TQM to thousands of companies. His approach is easy to understand, it is a methodical and a straightforward way of implementing a quality initiative. The tools he uses – quality process improvement teams, statistical analysis and process management (discussed later in this chapter) – are easily assimilated into FIT SIGMA.

Jan Carlzon

Sometimes, just a change in attitude and the recognition of key problem areas can be sufficient to make a big difference. For example, when Jan Carlzon took over Scandinavian Airlines (SAS) the airline was about to lose $US20 million. He found that SAS was a very efficient organization – it knew its business of transporting goods and people by air, and did this with clinical efficiency. It had sufficient resources and well-trained staff, and 10 million passengers were carried each year. Carlzon then established that for each passenger there were five occasions when the passenger came into contact with front-line employees, and that this contact lasted on average for 15 seconds. He called these contact times 'moments of truth' when he said (Carlzon, 1989):

> *Last year 10 million customers came into contact with approximately five SAS employees, and this contact lasted on average of 15 seconds each time. Thus, 'SAS' is created in the minds of the customers 50 million times a year, 15 seconds at a time. These 50 million 'moments of truth' are the moments that ultimately determine whether SAS will succeed or fail as a company. They are moments when we must prove to our customers that SAS is their best alternative.*

By establishing moments of truth, converting the staff to his way of thinking and taking some positive actions, within twelve months he was able to turn a $20 million loss into a $40 million profit.

However, this example is an exception. Few turnarounds are this dramatic, and generally benefits accrue over longer terms. The philosophy of TQM is to look for continuous improvement, not major breakthroughs; any major breakthrough is a bonus. No organization can ever say that TQM has been achieved – the quest for improvement is never-ending.

ISO 9000

In a discussion on the subject of quality it would be wrong to ignore the effect that the International Standard Organization 9000 series (ISO 9000) has had on quality. The ISO 9000 series and the more recent 14000 environmental series have been developed over a long period of time. The origins can be traced back to military requirements – for example, NATO in the late 1940s developed specifications and methods of production to ensure compatibility between Nato forces in weapons and weapons systems. In Britain ISO 9000's predecessor was the British standard BS 5750, which was introduced in 1979 to set standard specifications for military suppliers.

ISO 9000 certification means that an organization constantly meets rigorous standards (which are well documented) of management of quality of product and services. To retain certification the organization is audited annually by an outside accredited body. ISO 9000 on the letterhead of an organization demonstrates to its employees, to its customers and to other interested bodies that it has an effective quality assurance system in place.

Total Quality Management means more than just the basics as outlined in ISO 9000; indeed, ISO 9000 could be seen as running contrary to the philosophy of TQM. As Allan J. Sayle (1991) pointed out:

> *It is important to recognize the limitations of the ISO 9000 series. They are not and do not profess to be a panacea for the business's ills. Many companies have misguidedly expected that by adopting an ISO 9000 standard they will achieve success comparable to that of the over-publicized Japanese. One must not forget that the ISO 9000 standards did not exist when the Japanese quality performance improved so spectacularly: many Japanese firms did not need such written standards, and probably still don't.*

What does ISO 9000 achieve?

ISO 9000 exists primarily to give customers confidence that the product or service being provided will meet certain specified standards of performance, and that the product or service will always be consistent with those standards. Indeed, some customers will insist that suppliers are ISO accredited.

There are also internal benefits for organizations that seek ISO 9000 accreditation. First, by adopting ISO 9000 the methodology of the ISO system will show an organization how to go about establishing and documenting a quality improvement system. To achieve accreditation, an organization has to prove that every step of the process is documented and that the specifications and check procedures shown in the documentation are always complied with. The recording and documenting of each step is a long and tedious job; perhaps the most difficult stage is agreeing on what exactly the standard procedure is.

If an organization does not have a standard way of doing things, trying to document methods will prove difficult and many interesting facts will emerge. The act of recording exactly what is happening and then determining what the one set method should be is in itself a useful exercise. Non-value-adding activities will be unearthed and, hopefully, overall a more efficient method will emerge and be adopted as standard procedure. Determining a standard does not imply that the most efficient method is being used; the standard adopted only means that there is now a standard method (not necessarily the most efficient), that the method is recorded, and that the recorded method is used every time. The standard method not only includes the steps taken in the process, but also lists the checks and tests that are carried out as part of the process. This often requires the design of new and increased check procedures and a method for recording that each check or test has been done.

From this it can be seen that the adoption of ISO 9000 rather than streamlining an organization might actually serve to increase the need for audits and supervision. ISO 9000 can therefore, to this extent, be seen to be contrary to the philosophy of TQM. With TQM staff members are encouraged to do their own checking and to be responsible for getting it right first time, and the need for supervision becomes almost superfluous. With ISO 9000, the standard method will likely be set by management edict and, once set in place, the bureaucracy of agreeing and recording improvements may stultify creative improvements.

ISO tends to be driven from the top down and relies on documentation, checks and tests to achieve a standard, somewhat bland, level of quality assurance. TQM, on the other hand, once established relies on bottom-up initiatives to keep the impetus of continual improvement. However, as the Deming method of TQM does advocate a stable system from which to advance improvements, the adoption of the ISO 9000 approach means that there will be a standard and stable system. To this extent, ISO 9000 will prove a useful base for any organization from which to launch TQM.

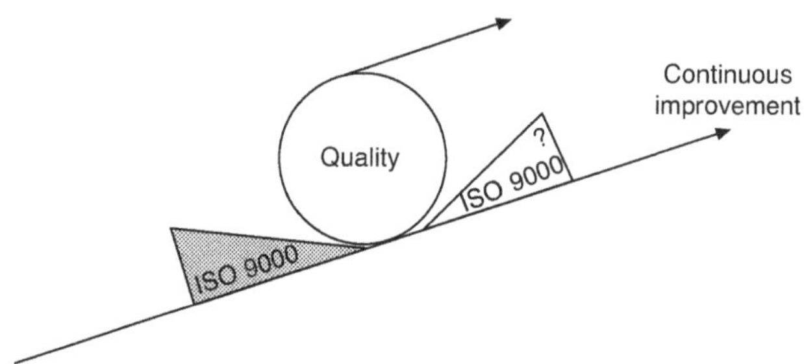

Figure 2.2 The wedge(s).

ISO 9000 – the wedge

As shown in Figure 2.2, ISO 9000 can be depicted as the wedge that prevents quality slipping backwards, but the danger is that it can also be the wedge that impedes progress.

Notwithstanding the benefits of obtaining a standard stable system through ISO procedures, it must be queried why a true quality company needs ISO 9000. If the customer or potential customer is not insisting on ISO accreditation, then the time and effort (and the effort expended will be a non-recoverable cost) makes the value of ISO to an organization highly questionable.

Gaining ISO 9000 accreditation is a long and expensive business. Internally it requires much time and effort, and most organizations underestimate the time and effort involved. Generally, recording the systems alone will require the full-time efforts of at least one person.

> *Example 2.1 ISO accreditation in a small print shop*
> One small print shop, employing twenty people and with one main customer, was sold the idea of ISO accreditation by a consultant and advised that the process of obtaining accreditation would take nine months. The actual time taken was two years and three months. The main customer had not asked for accreditation, but the difficulties experienced by the print shop in getting accredited led the customer to query the efficiency of the organization and the account was almost lost. What of the expensive consultant? Well, he took his fee and rode off into the sunset.

The internal costs of obtaining accreditation are expensive – more expensive than most organizations are prepared to admit. Total internal costs will not be known unless everyone involved in setting up the systems records and costs the time spent, and this is seldom done. The external costs can be equally expensive. It is not mandatory to hire an external consultant, but there are advantages in doing so. Consultants are not cheap, and quotations from at least three should be sought. Briefing the consultants will force an organization to do some preparatory work, which if properly approached should help in clarifying the overall purpose and give some indication of the effort that will be involved. Once the consultant is employed, it is the organization hiring the consultant that does the work. Consultants point the way – they give guidelines and hold meetings, they help with the planning – but don't expect them to get their hands dirty. They won't actually do any work; the organization seeking accreditation does the work!

Accreditation can only be obtained through an approved certifying body. The fee charged by the certifying body is relatively small, and depends on the size of the organization and the level of accreditation.

The ISO 9000 series has several standards – 9000, 9001, 9002, 9003, 9004 and, since 1996, the ISO 14000 series:

- ISO 9000 mainly deals with how to choose other ISO series standards for inclusion in a contract between a customer and a supplier.
- ISO 9001 should be chosen if there is design work or changes to designs involved, and/or if after-sales service is required.
- ISO 9002 should be chosen if there is no design work involved and/or no after-sales service in the contract. Some people think that ISO 9002 is easier to achieve and that therefore that ISO 9002 is a lesser 'qualification' to 9001. This is not so. If there is no design work involved or after-sales service required, then ISO 9002 is appropriate and it is no less onerous than 9001.
- ISO 9003 only requires one final check, and thus is not a good way of reducing costs of mistakes and of instilling a quality culture into the organization. Of course, ISO 9003 can be amended to include corrective action taken during the process and so on. If such amendments are made, then ISO 9002 may be more appropriate.

- ISO 9004 extensively uses the word 'should'. This means that an organization is not required actually to do anything included in the standard, and thus ISO 9004 can only be regarded as an advisory introduction paper to quality management. It is not so much that what ISO 9004 covers is wrong; it is the lack of compulsion that makes ISO 9004 of little value for contract purposes. If a customer were to use ISO 9004 in a contract document, then 'should' ought to be replaced with 'shall' throughout.
- The 14000 series concerns environmental (green) issues. Achievement of the standards is said to lead to business benefit through process performance improvement, cost reduction, reduced pollution, legislative compliance, and an improved public image. All very good – but if an organization has a social conscience and is environmentally aware, why would it need ISO accreditation?!

Throughout the ISO 9000 series, reference is made to documentation. To meet the ISO requirements, it is not necessary to have hard copies of quality plans, quality manuals and procedures. Indeed, when people have a computer terminal at hand they are more likely to search the computer than to leaf through large manuals. Also, with a computer system it is easier to update the records with the latest procedures and to ensure that the user acknowledges receipt of change when using the system. In this way the system can be kept almost instantly updated, and staff can be encouraged to make improvement suggestions.

The other important aspect of ISO is audits. The external audit requirements of the ISO 9000/14000 series are more towards compliance checks after an activity has started or been completed. This type of check confirms that procedures are being kept to, or that an outcome complies with the standard. Where mistakes are found, they are retrospective. Audits highlight where errors have occurred and thus indicate the need for corrective action for the future, but they don't stop the error happening in the first place. To be effective, where internal audits are in place the internal quality auditor should be trained in audit procedures and the purpose of auditing. Auditors should be there to help and guide, not to trap and catch. If the audit is preventative – that is, before the event rather than after – so much the better.

To summarize this discussion concerning ISO 9000, with TQM the aim is continuous improvement and, with the continuing impetus for quality, improvement being driven from the bottom up. ISO 9000 will not necessarily achieve this. At best ISO can be seen as a step on the way to TQM; at worst it might actually inhibit TQM, as it relies on the setting of top-down standards and controls and might deter staff from suggesting changes. A true TQM organization does not need ISO, but if ISO is insisted on by a customer it can be made to fit into the overall TQM plan.

Kaizen

The Japanese have a word for continuous improvement: kaizen. The word is derived from a philosophy of gradual day-by-day betterment of life and spiritual enlightenment. Kaizen has been adopted by Japanese business to denote gradual unending improvement for the organization. The philosophy is the doing of little things better to achieve a long-term objective. Kaizen is 'the single most important concept in Japanese management – the key to Japanese competitive success' (Imai, 1986).

Kaizen moves the organization's focus away from the bottom line, and the fitful starts and stops that come from major changes, towards continuous improvement of service. Japanese firms have for many years taken quality for granted. Kaizen is now so deeply ingrained that people do not even realize that they are thinking it. The philosophy is that not one day should go by without some kind of improvement being made somewhere in the company. The far-reaching nature of kaizen can now be seen in Japanese government and social programmes.

All this means trust. The managers have to stop being bosses and trust the staff; the staff must believe in the managers. This may require a major paradigm change for some people. The end goal is to gain a competitive edge by reducing costs and improving the quality of the service. In order to determine the level of quality to aim for, it is first necessary to find out what the customer wants and to be very mindful of what the competition is doing.

The daily aim should be accepted as being kaizen – that is, some improvement somewhere in the business.

Quality circles

In the 1960s Juran said (Juran, 1988):

> *The quality-circle movement is a tremendous one which no other country seems to be able to imitate. Through the development of this movement, Japan will be swept to world leadership in quality.*

Certainly Japan did make a rapid advance in quality standards from the 1960s onwards, and quality circles were part of this advance. However, quality circles were only one part of the Japanese quality revolution.

Quality circles have been tried in the USA and Europe, often with poor results. From our combined first-hand experience of quality circles in Australasia, the UK and Europe, South America, Africa, Asia and India, we believe that quality circles will work if the following rules are applied:

1. The circle should consist only of volunteers
2. The members of the circle should all be from different functional areas
3. The problem to be studied should be chosen by the team, and not imposed by management. Problems looked at by the circle may not always be directly related to quality or, initially, be seen as important by management.

4. Management must wholeheartedly support the circle, even where initially decisions and recommendations made by the circle are of an apparently trivial nature or could cost the company money (such as a recommendation for monogrammed overalls).
5. The members of the circle will need to be trained in working as a team (group dynamics), problem-solving techniques, and in how to present reports. The basic method study approach of asking why (what, where, when, who, and how) is a standard quality circle approach to problem solving, and members need to be taught how to apply this structured approach to solving problems.
6. The leader of the circle and the internal management of the circle should be decided by the members.
7. Management should provide a middle manager as mentor to the circle. The mentor's role is to assist when requested and generally to provide support. The mentor does not manage the circle.

The overall tenor of these rules is trust and empowerment. Management of the organization has to be seen to be willing to trust the members of the circle to act responsibly, and must then be active in supporting the circle. Although initially the circle may not appear to be addressing hard quality issues, very real benefits can be expected as the confidence of the members increases.

Side benefits of quality circles, which are nonetheless important, are the fostering of a supportive environment that encourages workers to become involved in increasing quality and productivity, and the development of the problem-solving and reporting skills of lower-level staff.

In Japan, the quality circle traditionally meets in its own time rather than during normal working hours. Not only do circles concern themselves with quality improvement; they also become a social group engaged in sporting and social activities. It is not expected in a European country that a quality circle would meet in the members' own time; few workers are that committed to an organization. However, there is no reason why, once the quality circle is up and running, management could not support and encourage social events for a circle, perhaps in recognition of an achievement.

Quality project teams

A problem experienced in the UK was the blurring of quality circles and quality project teams. The project team approach is top down – that is, management selects a hard quality problem and designates staff to be members of the team. The top-down, conscription approach might appear to be more focused than the quality circle approach, but the fundamental benefits of a voluntary team approach are lost. With the pure bottom-up quality circle approach, the members are volunteers and the circles consist of people who work well together and who want to contribute to the success of the organization.

Ishikawa (fishbone technique), or cause and effect

The Ishikawa diagram, named after its inventor Kaoru Ishikawa (1979, 1985), or the cause and effect diagram, is designed for group work. It is a useful method of identifying causes and provides a good reference point for brainstorming (brainstorming is discussed in a separate paragraph later in this chapter).

The usual approach is for the group to agree on a problem or effect. A diagram is then drawn consisting of a 'backbone' and four (or sometimes more) fish bones to identify likely causes. Common starting points are people, equipment, method, and material.

The following eight causes cover most situations:

1. Money (funding)
2. Method
3. Machines (equipment)
4. Material
5. Marketing
6. Measurements
7. Management and mystery (lack of communication, secret agendas etc.)
8. Maxims (rules and regulations).

Example 2.2 A large travel agency

Consider the situation where customers of a large international travel agency sometimes find when they arrive at their destination that the hotel has no knowledge of their booking.

In this case, to get started, the quality circle might begin with four basic possible causes: people, equipment, method and supplier (Figures 2.3, 2.4).

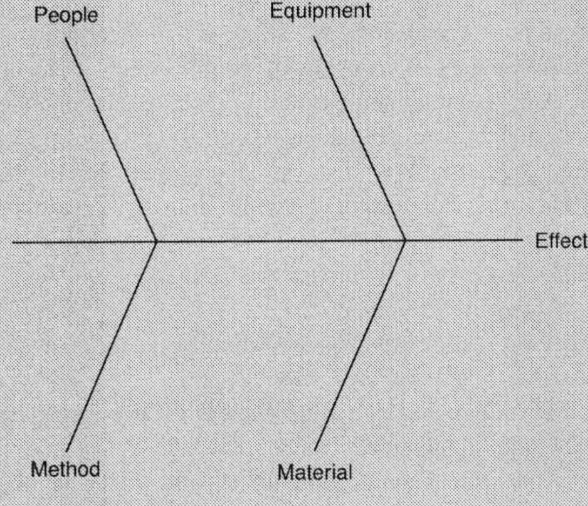

Figure 2.3 Ishikawa diagram.

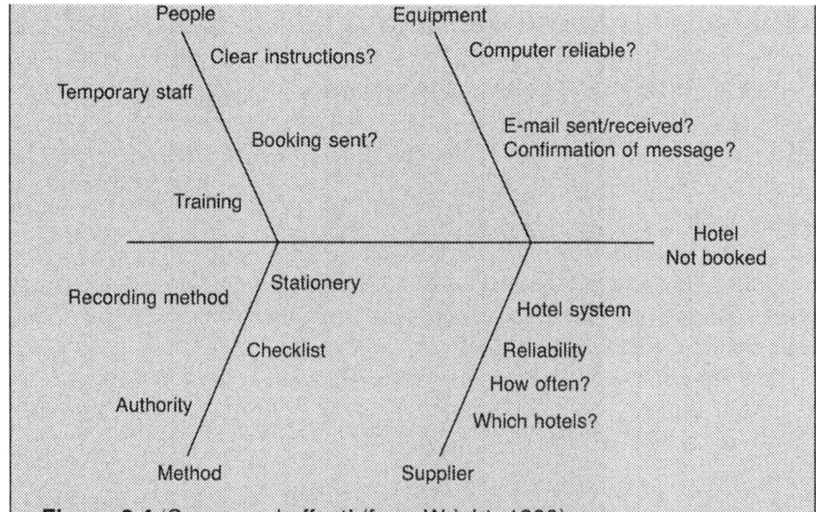

Figure 2.4 'Cause and effect' (from Wright, 1999).

The circle now has a clear picture of the possible problem areas and the linkages. The diagram points the way for collection of data. In this example, training and standard procedures appear to be worthwhile areas to follow up, and a second likely cause might lie with the suppliers (hotel systems); finally, the e-mail system might need checking. In a more detailed problem, sub-causes may need further breakdown until the true cause of the effect is determined. The main question is 'What and why?', see Table 5.1, Chapter 5.

Brainstorming

Brainstorming should be considered as a fun way of identifying all the causes of a problem. It consists of a group of people being given a problem to consider, with every person encouraged to make at least one suggestion.

Before the actual brainstorming process begins, it is important that the subject is defined and the rules of the session agreed. Members of the team will need at least five minutes of thinking time before the brainstorming proper begins.

Some rules for successful brainstorming are:

- Make one person responsible for recording suggestions on a white board or large flip chart.
- Encourage everyone in the team to 'freewheel'. There should be no criticism of seemingly silly suggestions.
- Every one in the team should come up with at least one suggestion, and other members of the team should not interrupt or make comments.

- Take suggestions by working around the room so that every one has a turn.
- If someone is unable to contribute first time round, pass on to the next person.
- Typically, there will be a lot of suggestions in the first 20 minutes, then there will be a lull. Don't stop when this lull occurs; keep going, as usually there will be another burst of ideas. Often the second burst provides the most creative ideas.
- Keep the initial ideas in front of the team until the end of the brainstorming session.
- When suggestions have dried up, review (as a team) the suggestions made and sort them into logical groups. Some suggestions will be found to be duplications and can be eliminated. One method of sorting the suggestions is to use a form of the cause-and-effect diagram.

Compatibility with FIT SIGMA

All of the foregoing methods are compatible with FIT SIGMA. Some of them, such as ISO 9000/14000, are not necessary, but if they exist in an organization then they are not wasted and provide a good foundation to move up to FIT SIGMA. FIT SIGMA is both a philosophy and an improvement process. The underlying philosophy is that of a total business-focused approach underpinned by continuous reviews and a knowledge-based culture to sustain a high level of performance. In order to implement the FIT SIGMA philosophy, a systematic process is recommended. This process is not a set of new or unknown tools; in fact these tools and this culture have been proven to yield excellent results in earlier waves of quality management. The differentiation of FIT SIGMA is the process of combining and retaining successes. Its strength is that it is not a rigid programme in search of problems, but an adaptable solution fit for any specific organization.

FIT SIGMA is a new and exciting approach to harnessing and sustaining gains from previous initiatives to secure operational excellence.

Summary

This chapter has covered the history of quality, reflected on the contributions of quality gurus since the 1960s, and considered some approaches and basic tools used in various quality systems.

Our belief is that quality is not a new or separate discipline; that it pervades all management actions. Our philosophy is that quality is too important to be left to the managers and that quality is everybody's concern – not only within the organization, but also of customers, suppliers and any other stakeholder.

Quality has two main aspects; it can be measured from the customers' perspective (customer satisfaction) and it can be viewed from the perspective of efficient use of resources. These two seemingly separate objectives are in fact inseparable when quality is considered. An organization that wishes to compete in the global market must be efficient and provide a high level of customer satisfaction. No organization will be able to afford to provide world-class service unless its use of resources is efficient and non-value-adding activities have been minimized, and no organization can afford not to be world class.

This chapter also briefly considered the part ISO 9000/14000 has in a total quality approach. Specific techniques such as quality circles and cause and effect analysis were also introduced. Finally, it was shown that elements of all the quality initiatives over the last 40 years are compatible with FIT SIGMA. FIT SIGMA is the new wave that enables an organization to maintain operational fitness.

3

The enigma of Six Sigma

Probable impossibilities are preferred to improbable possibilities.
Aristotle

Introduction

In this chapter we discuss the essence of Six Sigma; what it is and what it isn't. For a start, it isn't the new maddening management fad that keeps front-line workers too busy collecting data to do their jobs, *nor* is it an instant fix that will overcome corporate inefficiency and save trillions of dollars. Properly applied, Six Sigma has a proven record of creating a huge impact for organizations in performance. For this to happen, Six Sigma does require dedication and strong overt management support. It builds on the philosophy of Total Quality Management by adding advanced computer programs for analysis and benchmarking of performance. It requires an infrastructure of a small group of highly trained in-house technical consultants. It is a whole systems approach, and it cannot be applied haphazardly.

Challenges

For those championing Six Sigma there are many challenges, of which the two most significant are:

1. Initiative fatigue – the confusion and tiredness of staff when there is a succession of new initiatives to generate productivity and quality (e.g. a succession of maddening management fads that keep front-line staff from doing their jobs!).
2. The fear of statistical complexity – staff either see Six Sigma as 'too hard' or as statistical mumbo jumbo. Or because they can't understand, or are afraid that they will be shown up, they find it more comfortable to denigrate rather than to learn.

In this chapter we will take the mystery out of the apparent statistical complexity of Six Sigma, remove the enigma of Six Sigma and show that there is nothing to fear. The result will be a better product and service for the customer, and a simplified, foolproof, easy-to-control process for the organization – a process that is efficient and will save a fortune in reducing the cost of sales.

What is Six Sigma?

Sigma is a classical Greek letter (σ) that is used in mathematical and statistical models to signify the standard deviation from the mean. This might sound like statistical mumbo jumbo, but in reality is a very simple concept. The mean (more correctly referred to as the arithmetic mean) is what most of us would call the average – for example, if a cricket player bats in 10 matches and achieves a total score of 650, then the average is 65 (even though the player might have 'scored' nil on one occasion and 250 on another). Each turn at batting contributes to the average. In statistical terms, the arithmetic mean of the total score of 650 is 65 (650/10).

The next basic concept in statistics is frequency distribution. An often-used example in statistical textbooks is the tossing of ten coins 100 times. The result of each throw of the ten coins could range from ten heads and no tails to ten tails and no heads, with any combination in between (i.e. one head and nine tails, two heads and eight tails and so on). We would expect that, with evenly balanced coins, we would be more likely to get five heads and five tails than ten heads and no tails!

Table 3.1 gives the result of this experiment, which can be shown as a histogram (Figure 3.1) or as a distribution curve (Figure 3.2).

Table 3.1 Results of tossing 10 coins 100 times

Number of heads	0	1	2	3	4	5	6	7	8	9	10
Frequency	1	2	5	12	18	23	16	10	9	3	1

The curve shown in Figure 3.2 is an example of a normal distribution curve. The curve is bell-shaped (i.e. it is symmetrical from the midpoint). Of course not all distributions will give this shape, but under normal circumstances, given a large enough population (in our example 100 throws of 10 coins), it is very likely that the distribution curve will be similar to that shown in Figure 3.2. The midpoint is shown on our curve as 'x'. In statistical language, x represents the measure of central dispersion; in everyday English 'midpoint' indicates the same thing and is good enough for us!

If we assume a normal distribution curve as shown in Figure 3.2, one standard deviation from both sides of the midpoint (midpoint plus or minus one sigma) will include 68.27 per cent of the total; two standard deviations (two sigma) from both sides of the midpoint will include 95.45 per cent of the total; and three standard deviations (three sigma) will cover 99.73 per cent. If

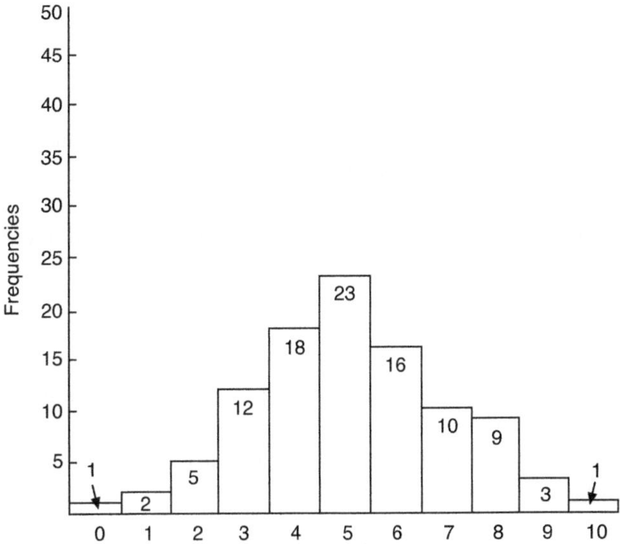

Figure 3.1 Histogram showing the results of tossing 10 coins 100 times.

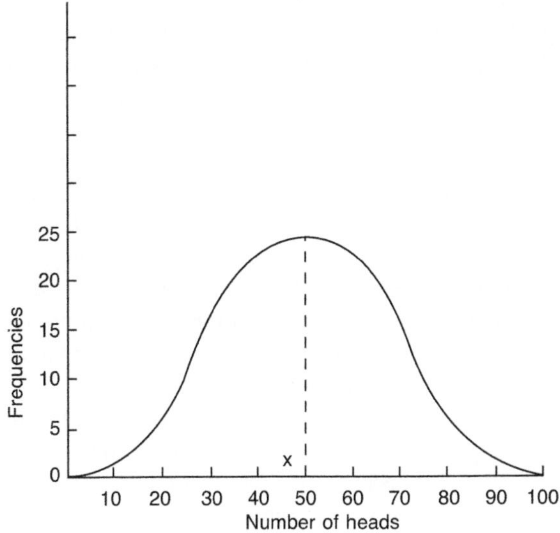

Figure 3.2 Distribution curve showing the results of tossing 10 coins 100 times.

we extend out to six standard deviations (six sigma) from each side of the midpoint, we cover 99.99966 per cent of the total!

For the quality programme known as Six Sigma, for a process, the higher the sigma the more of the outputs of the process (be they products or services), and the closer they will be to always meeting the customer requirements – or in other words, the higher the sigma the fewer the defects. It should be noted that the higher multiple sigma does not increase the 'variation'; it only increases the area covered under the distribution curve. For example:

- With one sigma, 68.27 per cent of products or services will meet customer requirements and there will be 317 300 defects per million opportunities
- With three sigma, 99.73 per cent of products or services will meet customer requirements and there will be 2700 defects per million opportunities
- With six sigma, 99.99966 per cent of products or services will meet customer requirements and there will be 3.4 defects per million opportunities.

Thus, in effect, Six Sigma is a theoretical statistical measurement allowing the measurement of the quality of products and services to a position where there are practically zero defects for any product or process in an organization.

Example 3.1 How to calculate sigma, or standard deviation

The mean deviation is the deviation of every item from the agreed average (see arithmetic mean above). If we sum the deviations and find the average value, we have the mean deviation. Because some items will be below the average (minus) and some will be above (plus), we will have negative and positive items. If we summed all the deviations and allowed for the plus and minus signs, it is possible that the total would be zero (pluses and minuses cancelling each other out). Thus we ignore the signs and treat all deviations as positive. The justification for this is that we only want to know the spread of the items around the mean; we are not concerned if they are above or below it.

Although ignoring the signs (plus or minus) is sufficient to show the distribution of data around a central position, disregarding the signs can be dangerous for advanced statistical work. Thus the mean deviation is seldom used, and the standard deviation is preferred.

The standard deviation is found in the same way as the mean deviation, but the plus/minus signs are not ignored. Instead we square the deviations, which makes all the items positive. For example, $+2 \times +2 = 4$ and $-2 \times -2 = 4$, and thus the negative signs are eliminated when data is squared.

The standard deviation for data is found in the following manner:

- Calculate the mean
- Find the deviations from the arithmetic mean
- Square each deviation (all items will now be positive)
- Add the squared deviations
- Find the average of the squared deviations, this is known as the *variance*
- Take the square root of the variance.

This can be expressed mathematically as

$$\sigma = \frac{\sqrt{\Sigma(\chi - \bar{\chi})2}}{n}$$

where σ = standard deviation; Σ = sum (or total) of; χ = each individual item of data; $\bar{\chi}$ = the arithmetic mean; and n = the total number of observations.

Consider a set of five observations: 20, 18, 22, 16, and 24. The total is 100, and the mean or average is:

$$\bar{\gamma} = \frac{\Sigma\gamma}{n} = \frac{100}{5} = 20$$

$\bar{\chi}$	γ	Dev	Dev2
20	20	0	0
20	18	-2	4
20	22	2	4
20	16	-4	16
20	24	4	16

Variance = 40/5 = 8

Standard deviation = √8 = 2.828

For this example, one standard deviation (2.828) either side of the mean equals 68.27 per cent of the total population.

Why Six Sigma?

Six Sigma is not just a statistical approach to measure variance; it is a process and culture to achieve operational excellence. Following its success, particularly in Japan, Total Quality Management (TQM) seemed to be everywhere. Sayings synonymous with TQM, such as work smarter not harder, right first time and every time, zero defects, quality is free, fitness for purpose, and the customer is king, became so over-used as to become hackneyed clichés. Although TQM was the 'in' management tool in the 1980s, by the 1990s it was regarded by many (especially in the USA) as an embarrassing failure and was written off as a concept that promised much but failed to deliver. Some believe that TQM swung too far towards the 'soft' issues of culture, consensus and staff involvement – the 'country club' approach. In any event, research by Turner (1993) has shown that any quality initiative needs to be reinvented at regular intervals to keep the enthusiasm level high. Against this background, Six Sigma emerged to replace the 'tired' TQM philosophy.

Six Sigma began in 1985 when Bill Smith, an engineer with Motorola, recommended the reinserting of hard-nosed statistics into the blurred philosophy of TQM. It should be remembered that Deming, Juran and Feigenbaum (see

Chapter 2), the pioneers of TQM, were statisticians and had all preached the need for measurement and control charts. Indeed, in Japan there was never a dilution of the statistical roots of TQM. Only in the USA and Europe, with the 'soft' politically correct approach to staff and the idea that if customer service is high ('have a nice day') the numbers (bottom line) will look after themselves, was the statistical approach to TQM lost. Although Six Sigma began in Motorola, its greatest successes have been in Allied Steel and General Electric. Following the recent merger of these two organizations, General Electric has become the worldwide leader for Six Sigma.

The key success factors differentiating Six Sigma from TQM are:

- The emphasis on statistical science and measurement
- Structured training plans at different levels (Champion, Master Black Belt, Black Belt and Green Belt)
- The project-focused approach with a single set of problem-solving techniques such as DMAIC (Define, Measure, Analyse, Improve and Control)
- The reinforcement of Juran's tenets, such as top management leadership, continuous education and annual savings plan
- That the effects are quantified in tangible savings (as opposed to TQM, where it was often said we can't measure the benefits, but if we didn't have TQM who knows what losses might have occurred?)

This last point, the quantification of tangible savings, is a major selling point for Six Sigma.

It is usually possible to measure the Cost of Poor Quality (COPQ) with the sigma level at which the process consistently performs. The COPQ if the performance level is at six sigma will be less than 1 per cent of cost of sales, whereas at three sigma (three sigma is regarded by many organizations as a very acceptable level of process quality) the corresponding COPQ will range from 25–30 per cent of cost of sales. These figures are explained and justified later in this book.

The structured approach of Six Sigma

Following the rigorous application of Six Sigma in many organizations, including Motorola, Allied Signal, General Electric, Bombardier, ABB, American Express, Wipro, GSK and others, a proven structured approach has emerged for product and process improvement. This structured and hierarchical process is shown in Figure 3.3.

The top management must have a total commitment to the implementation of Six Sigma and accomplish the following tasks:

1. Establish a Six Sigma leadership team
2. Develop and roll out a deployment plan for the training of Master Black Belts, Black Belts and Green Belts

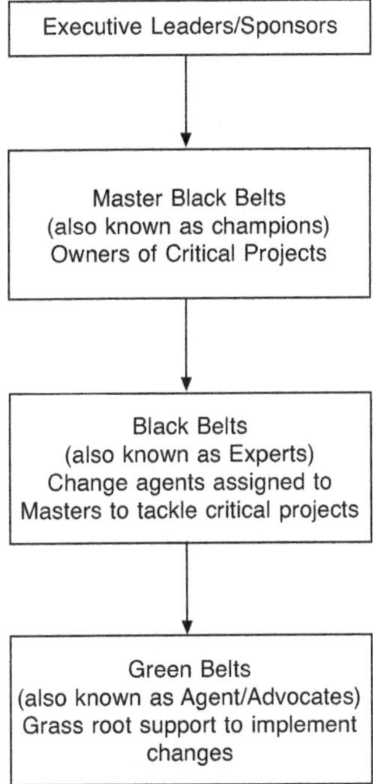

Figure 3.3 The structured approach of Six Sigma.

3. Assign Master Black Belts to identify and own critical projects related to key business issues
4. Provide support to Black Belts to make breakthrough improvements in critical projects
5. Encourage Green Belts to identify and implement 'just do it' projects
6. Set aggressive Six Sigma targets
7. Continuously evaluate the Six Sigma implementation and deployment programmes and make changes if necessary.

Six Sigma deployment

A critical piece of the successful Six Sigma experience is the Six Sigma deployment plan. A typical plan includes four parts:

1. Business alignment planning
2. The first wave of Black Belt training
3. The second wave (and subsequent waves as required) of Black and Green Belt training
4. Infrastructure development to deliver results and sustain culture.

Business alignment planning assures that Six Sigma projects align with business strategy and drive results. Preliminary project selection criteria start with projects having potential substantial savings (US$ 1 million or more). During the leadership education programme, which usually lasts from five days, Master Black Belts are selected and Masters are assigned to key business issues and projects. Intensive Master Black Belt training takes three to four weeks. While the Masters are being trained, the selection process for those to be trained as Black Belts is carried out.

Master Black Belts (MBBs) have a thorough understanding of the improvement process, DMAIC, Six Sigma-associated statistics and change management. Master Black Belts are capable of leading and managing significant end-to-end projects, and also coaching Green Belts.

Black Belt training is a combination of formal classroom training and on-site project work. Training is spread over four to five months, and includes four or five weeks of classroom work with the balance on-site project work.

Green Belts (GB) have an awareness of Six Sigma principles and in particular apply the DMAIC cycle to work with and support Black Belts in end-to-end projects. Green Belt training is usually spread over six to seven weeks, including five days of formal classroom work.

External consultants and experienced trainers are usually needed to manage and train the first deployment wave. For the second and third waves of the deployment plan, in-house Master Black Belts take charge of both the training and the management of specific projects.

Certification of Black Belts and Master Black Belts

Black Belts are awarded a certificate after:

- Completing three weeks of formal classroom Black Belt training
- Completing one week of change management and project management training
- Completing, as a team leader, an end-to-end $1 million Black Belt project
- Demonstrating Six Sigma commitment, evidenced by mentoring Green Belts and preparation of Green Belt training material.

Master Black Belts are awarded Masters' certificates after:

- Black Belt certification
- Completion of five end-to-end Black Belt projects as team leader
- Delivery of three Green Belt training sessions
- Demonstrating commitment to the Six Sigma philosophy through mentoring Black Belt and/or Green Belt projects
- Designing Black Belt training materials.

Sample of Black Belt training

(Minitab software is used in Statistical Applications.)

Week One	Week Two	Week Three	Week Four
Define and measure	Analyse	Improve and control	Advanced statistical tools
Variation	Process mapping	Generating improvement	Design of experiments
Sigma calculations	Root cause analysis	Selecting solutions	Process capability
Data collection	SPC tools	Change management	Process optimization
Sampling	Creativity	Key Performance Indicators (KPIs)	Multivariance charts
Cost of poor quality (non-conformance analysis costs)	Input/output	Project management	Design for Six Sigma

A sample deployment plan is shown in Figure 3.4.

World class

The success of Six Sigma has been well publicized for companies respected for their 'world class' engineering and manufacturing excellence. But what does world class signify?

As explained in Chapter 2, the term 'world class' has been attributed to Hayes and Wheelwright (1984), who related it to the capabilities of Japanese and German firms competing in export markets. Since then the term world class has been expanded to include such concepts as lean production, being the best competitor in at least one area of manufacture, growing more rapidly and profitably than the competition, hiring and retaining the best people, outpacing the competition in responding to market shifts (price changes, innovation of new services and products), and always continuously improving. Thus world class has come to mean a capability to deliver innovative and quality products at a lower cost, when the customer wants them, at a level better than the competition. Any definition of world class will now include the five aspects of quality, timeliness, flexibility, innovation and competitiveness (adapted from Fry and co-workers, 1994). The point is that what was best practice ten years ago is not what best practice is today. Nor does copying or chasing an organization deemed to have achieved world class mean that the chasing and imitating organization will also achieve world class. What is good for one organization at a point in time may not necessarily be good for another organization.

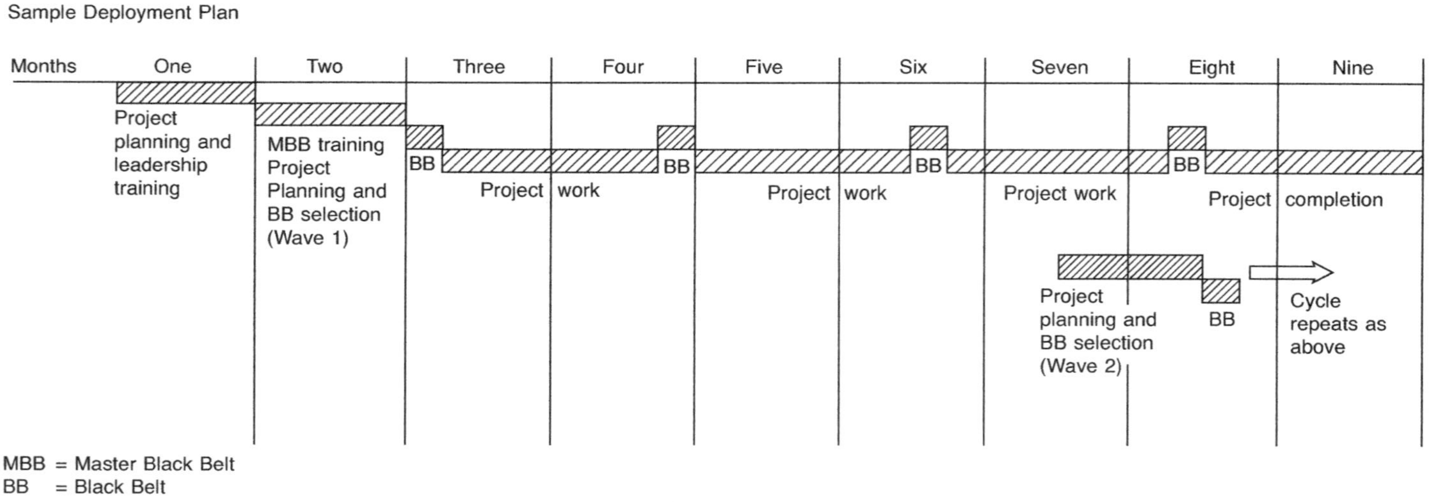

Sample Deployment Plan

Months One Two Three Four Five Six Seven Eight Nine

Project planning and leadership training

MBB training
Project Planning and BB selection (Wave 1)

BB

Project work

BB

Project work

BB

Project work

BB

Project completion

Project planning and BB selection (Wave 2)

BB

Cycle repeats as above

MBB = Master Black Belt
BB = Black Belt

Figure 3.4 Sample Six Sigma deployment plan.

Six Sigma in service industries

It might be considered that Six Sigma relates better to manufacturing, and, as stated above, the publicized successes have been in manufacturing and engineering operations. However, American Express, very much a service operation, began a successful Six Sigma pilot in 1998, and is now rolling out the Six Sigma programme worldwide. Additionally, much of the success of General Electric (GE) was in the GE Capital division (the financial services operation that accounts for over 40 per cent of GE's business). Bob Galvin, the former Chief Executive Officer of Motorola, has stated that the lack of initial Six Sigma emphasis in the non-manufacturing areas was a mistake that cost Motorola at least $5 billion over a four-year period.

One reason for this misconception is that service organizations are usually unaccustomed to looking at their processes in the traditional systematic 'input–process–output' manner of manufacturing (see Figure 3.5). A simplified input–

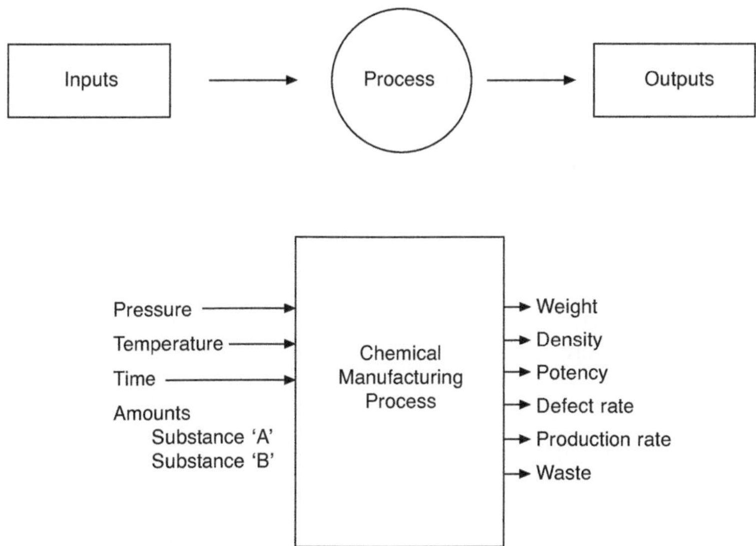

Figure 3.5 The Sigma input–process–output approach.

process–output (IPO) diagram for a sales process, as shown in Figure 3.6, illustrates the point that a process is a process regardless of the type of operation or organization. All processes have inputs and outputs; likewise, all processes have customers, suppliers and quality criteria.

The objective of Six Sigma is to gain significant breakthroughs and improved results by doing things better, faster and cheaper. Therefore, Six Sigma has to be incontrovertibly applicable to service industries.

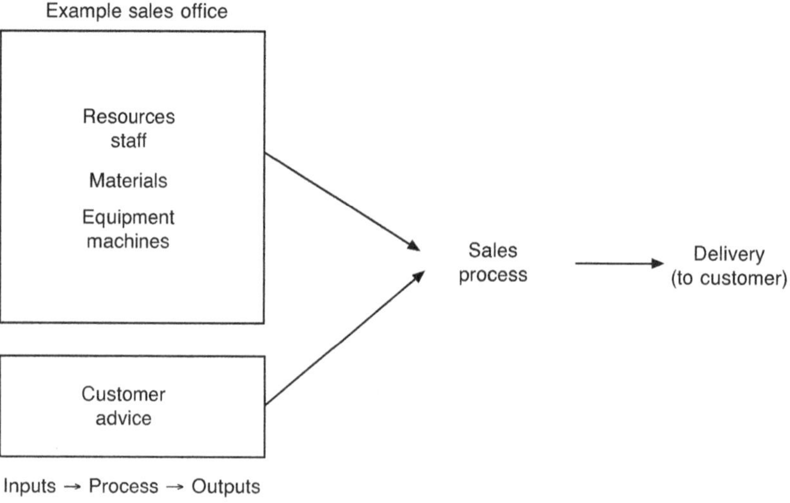

Figure 3.6 Simplified input–process–output diagram for a sales process.

Definitions and examples

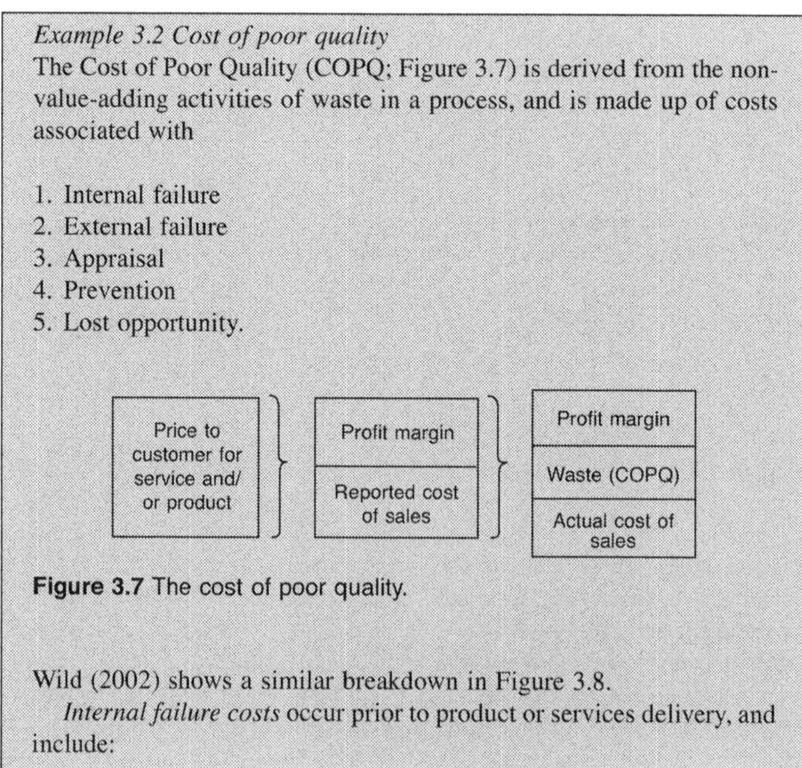

Example 3.2 Cost of poor quality

The Cost of Poor Quality (COPQ; Figure 3.7) is derived from the non-value-adding activities of waste in a process, and is made up of costs associated with

1. Internal failure
2. External failure
3. Appraisal
4. Prevention
5. Lost opportunity.

Figure 3.7 The cost of poor quality.

Wild (2002) shows a similar breakdown in Figure 3.8.

Internal failure costs occur prior to product or services delivery, and include:

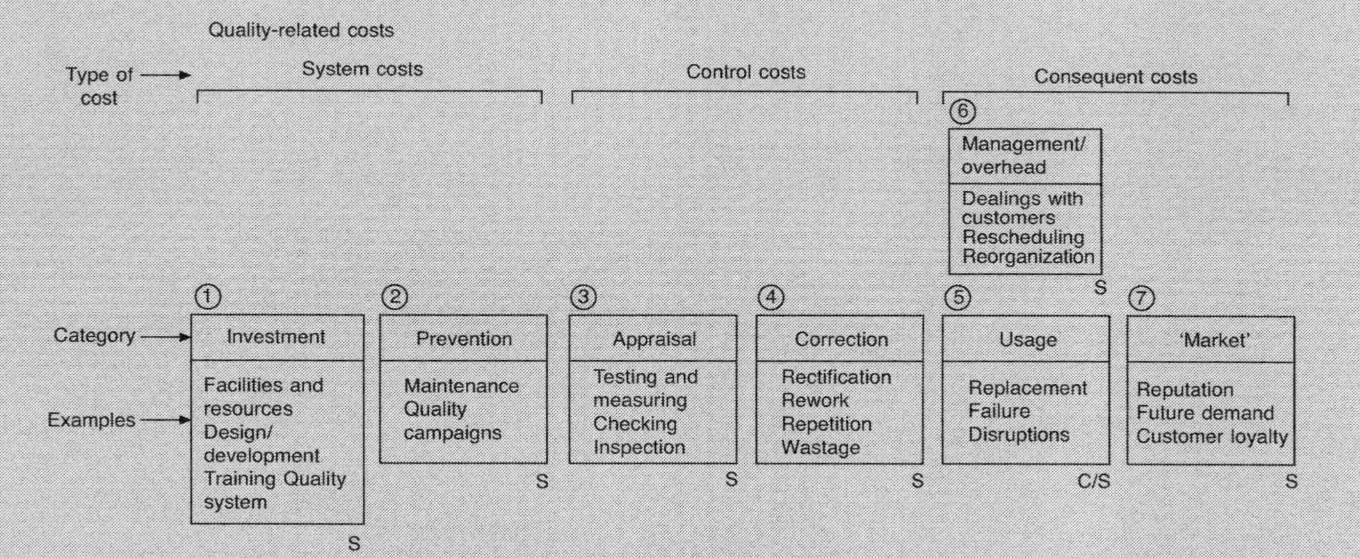

Figure 3.8 Quality-related costs (from Wild, 2002).

- Design corrective action
- Purchasing failure costs
- Production rework
- Material losses
- Energy losses
- Machine down time
- Overtime and labour costs
- Supervision and inspection costs
- Costs of management overheads.

External failure costs are incurred after product service provision, and include:

- Cost of recalls/withdrawals
- Returned/replacement goods
- Liability claims
- Customer complaint investigations
- Adverse regulatory reports/actions
- Loss of customer goodwill and loyalty
- Effects on reputation/future demand/marketplace perceptions.

Appraisal costs are associated with measuring, evaluating, testing and auditing products or services to ensure conformance with desired quality standards and performance requirements. These costs include:

- Purchasing appraisal costs
- Receiving and inspection of inwards goods
- Manufacturing appraisal costs
- Inspection resources (supervisors, inspectors and auditors)
- Laboratory support
- External appraisal costs, and cost of external auditors.

Prevention costs include all those costs associated with activities designed to ensure that products or services meet the needs of the customers. This includes not providing unwanted extras, and making sure that what is provided is what the customer actually wants. These costs include:

- Marketing research
- Customer perception surveys
- Supplier reviews and ratings
- Process validation
- Field trials
- Operator statistical process control (SPC)
- Quality education

- Continuous checks on workers and on processes to make sure that there is no deviation from the specification for goods or services.

Lost opportunity costs are associated with the longer-term negative impact on the business, and include:

- Lost sales
- Lost customers (both existing and potential new customers)
- Delayed market entry (and lost opportunities)
- Environmental and safety issues.

The cost of poor quality (COPQ) should be reported in both money ($) and in ratios. Each factor can be reported separately. Initial failure + External failure + Appraisal + Prevention + Lost opportunity costs is expressed as the Total Quality Cost in $.

$$\text{COPQ ratio} = \frac{\text{Total Cost of Quality \$}}{\text{Monthly Cost of Sales \$}}$$

Example 3.3 The define, measure, analyse, improve, control cycle

The Define, Measure, Analyse, Improve, Control (DMAIC) cycle is shown in Figure 3.9. DMAIC is the basic tool of the Six Sigma process. The five steps comprise:

1. Define opportunities (D). This is done through identifying, prioritizing and selecting the right projects. The key elements of this step include:
   - Validating business opportunities
   - Documenting and analysing possible projects
   - Establishing and defining customer requirements
   - Assessing benefits
   - Selecting projects.
2. Measure performance (M) of the projects and process parameters. The key elements of this step include:
   - Determining what to measure for inputs, process and outputs
   - Establishing a plan for data collection
   - Validating results and analysing variations
   - Determining the level of Sigma performance, allowing for process capability.
3. Analyse opportunities (A). Opportunities are analysed by identifying key causes and process determinants. The key features of this step include:
   - Analysing the input–process–output (IPO) to focus on problem areas
   - Analysing the flow process to identify non-value-adding activities
   - Determining root causes
   - Validating root causes.

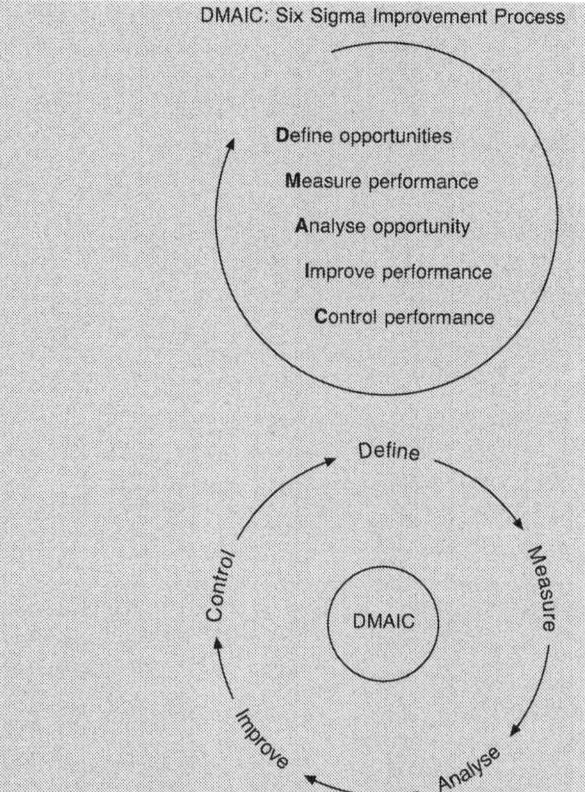

Figure 3.9 DMAIC cycle: Six Sigma improvement process.

4. Improve performance (I). This is achieved by changing the process so as to optimize performance. The key elements of this step are:
   - Generating improvement ideas
   - Quantifying and selecting solutions
   - Presenting recommendations
   - Implementing change.
5. Control performance (C). This is essential if gains are to be maintained. The key features of this step include:
   - Developing and executing pilot projects
   - Planning and implementing solutions
   - Monitoring and evaluating results
   - Project closure is established as standard operating procedure
   - Recognizing (celebration) of benefits gained.

Example 3.4 Design of experiments

Design of Experiments (DOE) is a technique of examining controlled changes of input factors and the observation of resulting changes in outputs – i.e. the response to input changes.

The origin of DOE is with R.A. Fisher in the 1920s, and his work in the Rothamsted laboratory for the agricultural industry (Fisher, 1925). The technique has been enhanced and applied in quality management, most notably by Ishikawa and Taguchi in Japan (see Chapter 2 for the Taguchi method).

The main objectives of this experimental approach are:

1. To obtain the maximum amount of information by using a minimum amount of resources
2. To determine which factors shift the average response, and which have little or no effect
3. To find settings for inputs that optimize the output and minimize the cost, and to validate results.

Figure 3.10 shows an example of an experiment designed to identify the effect of several factors.

Different strategies can be adopted for experimental designs, including:

- Adjusting one factor at a time

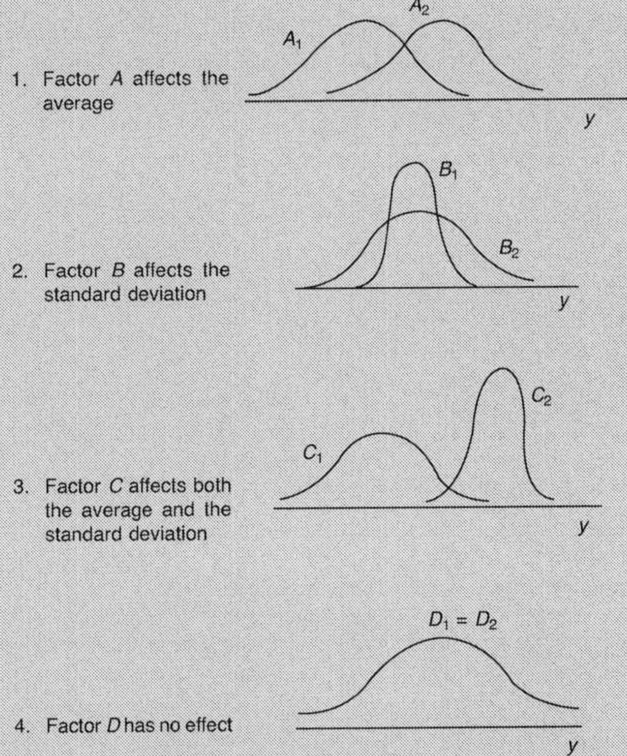

Figure 3.10 Example of an experiment to identify the effect of several factors.

- Adjusting several or all input factors
- Fractional adjustment of a factor
- The use of advanced models such as those developed by Taguchi.

The key steps for design of experiments are:

1. Define the problem and objectives – for example, test the different types of manufacturing processes for pharmaceutical tablets to find an optimal process.
2. Map the process with a simplified input–process–output (IPO) diagram – for example,

Input factors →	Process →	Outputs (responses)
Pressure →	Manufacture of tablet →	Hardness
Temperature		Weight
Time		Potency
Amount of substance 'A'		Defect rate
Amount of substance 'B'		Production rate

3. Select the best design strategy to suit your operation – for example, select the one factor at a time strategy.

P1	High pressure
P2	Low pressure
T1	High temperature
T2	Low temperature
H1	High process time
H2	Low process time
A1	High proportion of 'A'
A2	Low proportion of 'A'
B1	High proportion of 'B'
B2	Low proportion of 'B'

4. Conduct the experiment and record the data – for example,

Factors						Response values		
Run	P	T	H	A	B	y1... y12	$\bar{y}$	s
1	1	1	1	1	1			
2	2	1	1	1	1			
3	2	2	1	1	1			
4	2	2	2	1	1			
5	2	2	2	2	1			
6	2	2	2	2	2			

5. Analyse the data, draw conclusions, validate results and make predictions.

There are some proprietary software tools available for the analysis of data. MINITAB™ provides comprehensive functionality.

Example 3.5 Design for FIT SIGMA™

Design for FIT SIGMA (DFFS) utilizes the most powerful tools and methods known for developing optimal designs. These tools and methods include both DMAIC and DOE, and are capable of interfacing with any existing product development process of any organization. There are differences between DMAIC and DFFS, but the similarities are such that both approaches are compatible; one relates to process improvement and the other to product development and process. Both require a rigorous, disciplined approach.
The comparison is:

DMAIC	DFFS
Define	Define
Measure	Measure
Analyse	Analyse
Improve	Design
Control	Validate

In the context of DFFS, the first step, Define, has two objectives:

1. To get the project running, including the full involvement of the marketing function.
2. To agree and define critical quality factors as seen by the customer. In Six Sigma language this is referred to as Critical to Quality Factors (CTQ). It is important that the customers' requirements are understood when defining the CTQs. Six Sigma refers to this as Quality Function Deployment (QFD).

The second step of the DFFS project includes not only understanding customer needs but also measuring and prioritizing them. In many cases customer needs will be several, but not all will be of equal importance. The emphasis here is not on the measurement of process efficiency, but very much on what the customers (internal and external) expect from the outputs of the process.

In the analyse stage of DFFS, design options are examined and evaluated. The DOE technique is extensively applied to assess the response of each design factor and combination of factors. The next stage of DFFS focuses on design optimization, and establishes the best nominal settings of design parameters and tolerances. The final stage deals with

the validation of the design specifications. Verification ranges from ensuring that the design includes the establishment of process control once full-scale manufacturing begins.
DFFS provides many tangible benefits to organizations, including:

- Long-term cost reduction
- Reduction of time to market
- Definition of customers' needs, which improves quality in meeting these needs.

All of the above might sound straightforward, but experience has shown that there is a steep learning curve with the first project. There is also a need for a thorough understanding and application of the appropriate statistical tools. An extended version of DFFS, which has an even more in-depth customer focus, is known as Design for Customer Impact (DFC).

Summary

This chapter provides an introduction to the process and tools of Six Sigma. We have shown that the approach is to simplify and optimize the process to gain efficiency, and to make a determined effort to understand and measure exactly what customers really want. As Wild (2002) finds, the two key objectives of organizations are customer service and efficient use of resources, and these objectives are often in conflict. The aim of Six Sigma is to reduce the conflict by fully understanding what the customer really wants and then designing the product and process to minimize the costs of production.

This chapter has introduced several acronyms, but acronyms alone don't make the difference; understanding the philosophy and the rigorous adherence to the steps of Six Sigma will make the difference!

4

Case studies: Six Sigma in practice

If the cap fits, wear it.

Charles Dickens

It is well accepted that in the 1970s Japanese firms set the pace by their focus on quality and performance reviews. By comparison western companies (i.e. USA and European owned and operated companies) appeared to be less concerned about the importance of quality measures and the impact of standards of performance on bottom-line results. TQM initiatives that were taken seemed to concentrate on softer cultural issues rather than 'hard' performance standards. However, with the introduction of Six Sigma tools in the 1980s Motorola revolutionized the quality movement. Western companies that operated at levels of two to three sigma (with between 45 500 and 2700 defects per million operations) became increasingly interested in improving their performance standards (and their share price). Although 99 per cent sounds very good, it slowly dawned on companies that there is a tremendous difference between 99 per cent and 99.9997 per cent – for example, for every million articles of mail the difference is between 10 000 lost items and 3 lost items. Again, for 2 million prescriptions of medicine per annum 99 per cent = 20 000 wrong prescriptions, whereas 99.9997 per cent in theory equates to 7 wrong prescriptions (and for 20 million prescriptions the decrease is from 200 000 errors to 68). In practice 99.9997 per cent would mean no wrong prescriptions, as the process and culture is conditioned for zero defects rather than being one that accepts that it is inevitable, and acceptable, that mistakes will occur.

The first wave of organizations to use Six Sigma, following the grand groundwork of Motorola, included Allied Signal, Texas Instruments, Ratheon and Polaroid (to name but a few). GE entered the arena in the mid-1990s and in turn was followed by many powerful corporations, including SONY, HOB, Dow, Bombardier, and GSK.

The ability to leverage the experience of successful Six Sigma players proved highly attractive, both as a competitive issue and also to improve profit margins. The following case examples provide insights into organizations that are achieving success in their business performance through the use of Six Sigma programmes.

Case study 1: General Electric (GE)

With over 4000 Black Belts and 10 000 Green Belts across its businesses, and Six Sigma savings of $2 billion in 1999 alone, GE is a comprehensive Six Sigma organization. GE is the benchmark for Six Sigma programmes.

The company

General Electric has been at the top of the list of Fortune 500's most admired companies for the last five years, and without doubt their Six Sigma programme has played a key role in their continued success. In 2001 GE's turnover was over $125.8 billion, they employed 310 000 people worldwide, and their market value was $401 billion. With earning growing at 10 per cent per annum, GE also has the enviable record of pleasing Wall Street and financial analysts year after year. GE's products and business categories span a wide spectrum and include automotive, construction, health-care, retail, transport, utilities, telecommunications and finance industries.

Driver for Six Sigma

The CEO of GE, Jack Welch, is reported to have become attracted to the systematic and statistical method of Six Sigma in the mid-1990s. He was ultimately convinced of the power of Six Sigma after a presentation by Allied Signal's former CEO, Larry Bossidy, to a group of GE employees. Bossidy, a former Vice Chairman of GE, had witnessed excellent returns from Allied Signal's experience with Six Sigma.

In 1995 GE retained the Six Sigma Academy, an organization started by two early pioneers of the process, both ex-Motorola, Michael Harry and Richard Schroeder. It was pointed out that the gap between three sigma and six sigma was costing GE between $7 and $10 billion annually in scrap, rework, transactional errors and lost productivity. With the full and energetic support of Jack Welch, senior management became fully committed to the Six Sigma programme. 'GE QUALITY 2000' became the GE mantra for the 1990s and beyond. Jack Welch declared that 'Six Sigma, GE Quality 2000 will be the biggest, the most personally rewarding and in the end the most profitable undertaking in our history!'. While financial benefits and the share price were a driving force in Six Sigma deployment, GE identified four specific reasons for implementing it:

1. Cost reduction
2. Customer satisfaction improvement
3. Wall Street recognition
4. Corporate synergies.

Improvement programme

Although Motorola pioneered the Six Sigma programme in the 1980s to improve manufacturing quality and eliminate waste in production, GE broke the mould of Motorola's original process by applying the Six Sigma standards to its service-oriented businesses – GE Capital Services and GE Medical systems. Note that GE Capital Services accounts for nearly half of GE's total sales.

The Six Sigma programme was launched in 1995 with 200 separate projects supported by a massive training effort. In the following two years a further 9000 projects were successfully undertaken, and the reported savings were $600 million. The training investment for the first five years of the programme was close to $1 billion. GE also instituted a personnel recruitment plan to augment the cadre of dedicated full-time Six Sigma staff. Figure 4.1 shows the structure of a typical GE Six Sigma team.

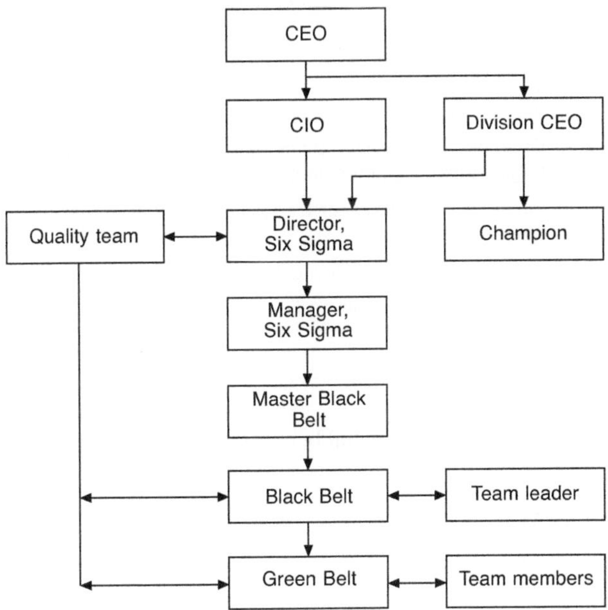

Figure 4.1 The structure of a typical GE Six Sigma team.

The GE programme revolved around the following few key concepts, all focused on the customer and internal processes:

- Critical to Quality – the determination of and development of attributes most important to the customer
- Defect – the identification of failure to meet customer wants
- Process capability – what the process can deliver
- Variation – what the customer sees and feels, as against what the customer wants

- Stable operation – ensuring consistent and predicable processes to improve what the customer sees and feels
- Design for Six Sigma – designing to meet customer needs and process capability.

Model for roll out

There does not appear to one universal model for roll out of Six Sigma amongst the companies that have implemented a Six Sigma programme. However, the Six Sigma Academy advises that there is a general model that is effective and has been adopted/developed by GE. This general model is shown in Table 4.1.

Table 4.1 GE training model

Phase one	Business units select champions and Master Black Belts. The Six Sigma Academy recommends one Champion per business group and one Master Black Belt for every 30 Black Belts
Phase two	Champions and Master Black Belts undergo training. The overriding deployment plan is developed
Phase three	Champions and Master Black Belts, with the assistance of Black Belts, begin identifying potential projects
Phase four	Master Black Belts receive additional training, focusing on how to train other staff
Phase five	Black Belts undergo training and the first projects are officially launched
Phase six	Black Belts begin training Green Belts

Key benefits achieved

Figure 4.2 shows the direct financial benefits achieved by GE over a four-year

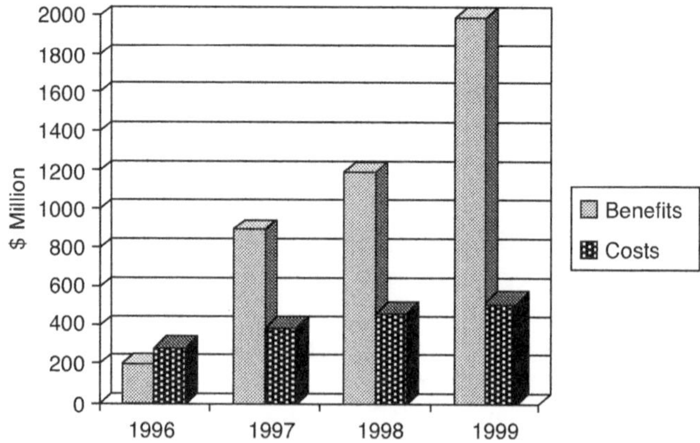

Figure 4.2 Six Sigma payoff at GE.

period. This provides evidence of the very real benefits that are achievable from a Six Sigma programme. With FIT SIGMA™, the next stage is to sustain and to grow the benefits.

At the second level of benefits, where the impact on savings is not direct, the achievements, average per year, include:

- 20 per cent margin improvement
- 12–18 per cent capacity increase
- 12 per cent reduction in headcount
- 10 per cent to 30 per cent capital expenditure reduction.

Some specific examples from business units are:

- GE Medical Systems – in the introductory year there were 200 successful projects
- GE Capital – invested $6 million over four years to train just 5 per cent of the work force who worked full time on quality projects, and 28 000 quality projects were successfully completed
- GE Aircraft Engines – the time taken to overhaul engines reduced by an average of 65 days
- GE Plastics – in just one project, a European polycarbonate unit increased capacity by 30 per cent in eight months.

Market consultants and analysts have reacted favourably to GE's achievements with Six Sigma. Merrill Lynch is quoted as saying: 'Six Sigma balance sheet discipline plus service and global growth are helping fuel (GE's) 13 per cent earning per share gains'.

On 8 May 2002 GE announced that it will deliver record earnings in 2002 of more than $16.5 billion, and it comfortably forecast double-digit earnings growth for 2003.

Lessons learned

At one level, to emulate GE maybe considered as being beyond the reach of many companies. It is cash rich, and its business generates over 10 billion dollars per month ($125.8 billion sales for 2001). It makes real things like turbines and refrigerators, and people buy their products with real money. However, on closer examination there are some strong learning points from the GE Six Sigma programme that can benefit *any* company embarking on a quality programme. These include:

1. *Leadership support.* There is absolutely no doubt from published data that the chief architect of success was Jack Welch. For any organization wanting to change a culture such as is required for Six Sigma and FIT SIGMA, strong unstinting leadership from the top is essential. Likewise, all senior management must be engaged and believe in the philosophy.

2. *Definition of Six Sigma objectives.* The objective is to be world class. World-class companies such as GE recognize that quality initiatives are synonymous with profit enhancement and share price. World class means internal efficiency, best practice and a focus of customer satisfaction. The lesson is that anything less than an ambition to be world class simply won't do.
3. *Development of initial processes and tools.* At GE each problem was defined through measurement and analysis along a five-step DMAIC (Define, Measure, Analyse, Improve and Control) approach, and the use of the seven quality 'tools' of control charts, defect measurement, Pareto analysis, process mapping, root cause analysis, statistical process control, and decision tree diagram. The lesson is that a structured approach has to be followed for Six Sigma process management.
4. *Alignment of Six Sigma with career paths.* At GE, Black Belt status became essential for staff on the fast track for advancement. Black Belts were rewarded with share options (in most companies share options are reserved for senior management). The lesson from this is that recognition has to be given to motivate and retain valuable talent.
5. *Six Sigma and service industries.* The piloting of Six Sigma in GE Medical Systems and GE Capital Services incontrovertibly proved that Six Sigma is not just for manufacturing; the process is equally applicable to all operations, including services. GE has opened the gate for service operations. In western economies, 80 per cent of gross domestic product is from the service sector.

Case study 2: The Dow Chemical Company

The company

The Dow Chemical Company, at just over 100 years old, is widely recognized as a technology-based manufacturing business. With annual sales of US$28 billion, Dow is the world leader in the production of plastics, chemicals, hydrocarbons, and herbicides and pesticides. Dow is also a leader on performance plastics (adhesives, sealants and coatings). Other products include polyethylene resins for packaging, fibres and films, as well as performance chemicals such as acrylic acid. Dow has recently bolstered its polyethylene operations with the acquisition of Union Carbide, and also produces commodity chemicals (chlorine and caustic soda) and oil-based raw materials. They have customers in more than 170 countries who have a wide range of markets, including food, transportation, health and medicine, personal and home care, and building and construction, among others. Dow has a policy of sustainable development, and uses 'triple bottom-line' results – an approach that measures success by economic prosperity, environmental stewardship and corporate social responsibility. The company has approximately 50 000 employees around the world, 208 manufacturing sites in 38 countries, and supplies more than 3200 products.

Drivers for change

Dow's mission is 'To constantly improve what is essential to human progress by mastering science and technology'. Dow has set its aspirations purposely high. These higher aspirations have fuelled the company's journey toward Six Sigma and business excellence. This mission is founded upon a long history of continuous improvement and corporate reinvention. Throughout the early 1990s, Dow employed a number of measures to streamline and improve its competitive position. Value-based management tools were instituted, quality performance mechanisms put in place and re-engineering practices implemented. In 1994, the company refocused and re-shaped its strategy. The result of this effort was a strategic blueprint containing four critical and interrelated components:

1. Competitive standard
2. Value growth
3. Culture
4. Productivity.

Following the development and implementation of the strategic blueprint, Dow continued its improvement journey. Global workstations established a communications pipeline that allowed all employees around the world to share a common computer systems, thereby accelerating the pace and quality of communications. Through this period Dow also implemented a people success system for the development and growth of human resources, and established a leadership development network to build on leadership skills and align the organization. The company also instituted growth acceleration initiatives to place increased focus on value growth, and launched strategic performance measures to track company performance against key metrics.

While the productivity measures implemented in the 1990s established strong competitive advantages, Dow leadership's vision extended beyond the role of leadership in the chemical industry and extended to business excellence. In late 1998, the Dow leadership embarked on a search for an enabler that would drive the company to the next level of productivity, performance and value. Leadership teams visited a number of top-tier global companies, holding discussions on the latest ideas and trends in productivity and improvement. The search led to Six Sigma.

Dow's implementation of Six Sigma began by taking a four-month hiatus to formulate a breakthrough implementation strategy. Within the context of this planning, a number of key decisions were made that set Dow's implementation of Six Sigma apart from that of others. One decision was that Six Sigma at Dow would be integrated into the business strategies of the company rather than being relegated to a corporate role. Many quality programmes of the past that were relegated to corporate roles were plagued with responsibility but little authority. In effect, this decision added vigour to Dow's implementation of Six Sigma by placing accountability for results directly on the shoulders of the business leaders of the company.

Additionally, Dow wanted to distinguish its practice of Six Sigma beyond a focus of the MAIC (Measure, Analyse, Improve, Control) methodology of Six Sigma by incorporating linkages to those strategic drivers that are at the centre of focus for the company. The first of those drivers is a concentrated emphasis on Six Sigma projects that drive customer loyalty. Second, Dow chose to create a Six Sigma linkage to the technology of leveraging. Throughout the 1990s, Dow instituted a global business model and a single information technology platform. With Dow's integrated business structure, single information systems platform and global technology centres, the company was in a strong position to leverage best practices from Six Sigma.

Implementation design

The Dow leadership team travelled to Scottsdale, Arizona, to meet with the Six Sigma Academy in early 1999. Following a series of meetings, two businesses within the company implemented Six Sigma. Late in the summer of 1999 Dow leadership made a bold commitment, expanding the implementation to all its businesses and functions around the world.

Under the leadership of Kathleen Bader, the 'Staircase of Change Leadership' (see Figure 4.3) was employed to develop an implementation designed to drive change in a revolutionary, yet sustainable, manner. Each successive step in this staircase builds upon the previous step, forming a solid foundation for change leadership. The steps in this staircase include:

1. *Vision.* Dow's stated vision for Six Sigma is: 'Dow will become recognized and lauded as one of the premier companies of the 21st century, driven by an insatiable desire to achieve a Six Sigma level of performance and

Figure 4.3 The Dow Staircase of Change Leadership.

excellence in all that we do'. Additionally, the Dow vision for Six Sigma was cast in the company's 1999 annual report to shareholders, 'delivering $1.5 billion in EBIT cumulatively by 2003 . . .'.

2. *Values.* Dow widely communicates its corporate values – integrity, respect for people, unity, outside-in focus, agility and innovation – to encourage all employees to honour the relationships.
3. *Attitude*: In its highest form, Six Sigma represents a mindset change that focuses on results, accountability, and data-driven decision making. In the environment of a large global corporation the unified, passionate attitude of leadership is essential to effective change. According to Kathleen Bader, 'It is an attitude that imposes accountability . . . and induces results'.
4. *Language.* The soul of attitude is evidenced in language. The implementation of Six Sigma utilizes its own terminology. Utilizing the common language of Six Sigma was instituted as a leadership practice.
5. *Behaviours.* A listing of behaviours was communicated throughout the company in 'road shows'. These behaviours included: adopting intolerance for variation, measuring inputs not just outputs, demanding measurement and accountability, requiring sustainable gains, delivering on customer satisfaction to build loyalty, and leveraging for competitive advantage.
6. *Best practices.* Dow undertook a diligent study of best-in-class Six Sigma practitioners in order to identify key success factors and gaps. From this study came numerous best practices. Additionally, gaps were identified that were employed to differentiate Dow's implementation of Six Sigma. Specifically, these gaps were customer loyalty and leveraging. The application of customer loyalty to Dow's implementation of Six Sigma is much more than lip service and good intention; up to 25 per cent of all Six Sigma projects are focused on driving a customer loyalty differential for Dow. Moving a customer from being satisfied to being loyal can create a powerful, sustainable business impact, and Dow has implemented a process model that drives this critical transformation. Leveraging is defined as the effective multiple implementation of demonstrated best practices. Breakthrough quality, coupled with Dow's unique ability instantaneously to transmit a Six Sigma solution from Texas to Taiwan, turns ideas into impact on the bottom line. Breaking down silos and unleashing the power of leveraging across every Dow business around the world is having a multiplier effect on the company's implementation of Six Sigma. Leveraging is an integral component of Black Belt training at Dow. Furthermore, Dow has established Leveraging Champions within each of its businesses.
7. *Articulated strategy.* The drivers for change facing Dow created an urgency that would not wait for evolution, and a detailed and rigorous breakthrough strategy was developed. The Six Sigma breakthrough strategy implemented at Dow wove together three leading edge processes:
   - The stages of change
   - The management of change
   - Managing implementation.

8. *Implementation*. Full-scale implementation of Six Sigma at Dow began early in 2000. As many as four training waves, each containing approximately 200 Black Belts, have been conducted since the full-scale launch. Champions and Process Owners are also identified to make sure that control plans stay in place and gains are sustained for the long term.
9. *Culture change*. A Six Sigma resource commitment was established by the company. This commitment calls for 3 per cent of all employees to be Six Sigma Black Belts. Black Belts are expected to fulfil a two-year, full-time commitment to Six Sigma. In addition, employee compensation plans are tied to Six Sigma results. Top leadership has established an expectation that all employees have at least one personal goal tied to Six Sigma, and all of its professional-level employees must be engaged in a successful Six Sigma project by year-end 2005.
10. *Success*. There is an old maxim: 'Nothing succeeds like success'. Dow's Six Sigma implementation is generating significant financial results, and is effectively driving positive, powerful cultural change.

Key benefits achieved

While Dow does not release its aggregate Six Sigma results, it has announced that by year-end 2001 the company was more than halfway towards its goal of achieving US$1.5 billion in cumulative Earnings before Interest and Taxes (EBIT) from Six Sigma.

Of all Six Sigma projects that have been closed through the realization phase, the average financial impact is US$520 000 per project.

At the time of this writing:

- Dow has 1269 active Black Belts. This represents 2.4 per cent of the company's current population. Although this is short of the 3 per cent goal, the employee population has grown significantly in recent times due to major acquisitions.
- 23 per cent of all Dow employees have been involved in a successful Six Sigma project.
- Dow has more than 2800 active Six Sigma (MAIC) projects and more than 100 active Design for Six Sigma projects.
- Despite challenging economic conditions, Dow fully expects to achieve its goal of US$1.5 billion in cumulative EBIT by 2003.

Lessons learned

1. *The value of constancy of purpose*. Dow began its focus on Six Sigma with top-down leadership endorsement. The power of that endorsement has been sustained and has grown since its implementation began. This 'constancy of purpose' sends a clear signal to the entire company about long-term expectations and true cultural change.

2. *Financial rigour.* Dow instituted business rules and established a team of trained financial analysts to review and validate financial benefits from its Six Sigma projects. Applying financial rigour to projects offers transparency and credibility to the company's implementation of Six Sigma.
3. *Data capture and knowledge management.* Six Sigma drives a data-based decision-making process. In order to capture and leverage knowledge, a flexible and user-friendly database must be established. Dow has invested significantly in the construction and maintenance of its database system for Six Sigma. This investment is paying substantial dividends in terms of knowledge capture for leveraging and tracking of project metrics for ongoing improvement.
4. *A way to do work . . . not an additive.* Many falsely believe that Six Sigma is additive or parallel – in other words, Six Sigma is often viewed as something else that the organization has to do rather than the way in which work is done. It is essential clearly to position Six Sigma as the way in which work is done
5. *Pipeline momentum.* One challenge of implementation through rapid transformation is that it is possible to deplete the project pipeline. Keeping a robust pipeline is essential to maintaining and building momentum for Six Sigma implementation. Time spent up front in creating a project pipeline is well spent.

(Sources: Kathleen Bader and Jeff Schatzer of The Dow Chemical Company, May 2002; http://www.moneycentral.msn.com 17 May 2002.)

Case study 3: Seagate Technology

The company

Seagate's position as the world's largest manufacturer of disk drives, magnetic disks and read–write heads and a leader in Storage Area Network (SAN) solutions puts it at the heart of today's 'information-centric' world. Since its founding in 1979, Seagate has successfully relied on a strategy of vertical integration – designing, developing and producing the key enabling technologies that go into its storage products, rather than relying solely on outside suppliers.

At the core of Seagate's success is its advanced development of hard disk drive products. Seagate is the market leader in each of the segments in which it competes, ranging from price-sensitive desktops to performance-intensive network servers, and produces a broad range of disk drives in capacities ranging from 20 Gbytes to an industry-leading 180 Gbytes. In the growing market of consumer electronics devices, which includes personal video recorder (PVR) products, gaming consoles and digital audio jukeboxes, Seagate has shipped nearly three million disk drives and taken market leadership.

Seagate Technology is a global company employing nearly 50 000 people, with R&D and product sites in Silicon Valley, California; Pittsburgh, Pennsylvania; Longmont, Colorado; Bloomington and Shakopee, Minnesota; Oklahoma City, Oklahoma; Springtown, Northern Ireland; and Singapore. Manufacturing and customer service sites are located in California, Colorado, Minnesota, Oklahoma, Northern Ireland, China, Indonesia, Malaysia, Mexico, Singapore and Thailand.

Drivers for change

Seagate is the world's leading provider of storage technology for Internet, business and consumer applications. Seagate's market leadership is based on delivering award-winning products, customer support and reliability to meet the world's growing demand for information storage.

Six corporate objectives drive all day-to-day activities within the company. They are:

1. Improve Time-To-Market (TTM) for all products
2. Lead the industry in key technologies
3. Create world-class manufacturing operations
4. Develop strategic relationships with vendors and key customers
5. Provide best-in-class product and process quality
6. Become an employer of choice.

Why implement Six Sigma? The market leadership of the company is continuously challenged in a highly competitive and dynamic environment, as is indicated by the following measures:

- Volume products remain in production for only 6–9 months
- Technology content doubles every 12 months
- Worldwide shipments of hard disk drives increases by 10–20 per cent per year
- Cost per unit of storage drops 1 per cent every year.

In 1998 Seagate's senior executive team was concerned that business performance was not on a par with expectations and capabilities. The quality group was charged with recommending a new model or system with which to run the business. The Six Sigma methodology was selected and launched in 1998 to bring common tools, processes, language and statistical methodologies to Seagate as a means to design and develop robust products and processes. Six Sigma helps Seagate make data-based decisions that maximize customer and shareholder value, thus improving quality and customer satisfaction while providing bottom line savings.

Six Sigma was one of the three key activities seen as essential for Seagate's continuing prosperity. The other two were:

1. Supply chain – how to respond to demand changes in a timely manner, execute to commitments and provide flexibility to customers
2. Core teams – how to manage product development from research to volume production.

For example, Seagate's lean manufacturing activities are a key part of Seagate's supply chain improvements and are increasingly tightly bound with Six Sigma. Lean manufacturing's value stream mapping approach and Six Sigma's analytical strength fit together extremely well to define, solve and then prevent problems.

As Six Sigma matures at Seagate, leaders are shifting their focus from reactive to proactive deployment and are placing further emphasis on weaving Six Sigma into business areas in addition to operations. The company also deployed Design for Six Sigma methodology (DFSS), providing new tools and an emphasis on designing products based on a systems engineering approach, so core teams are now starting to manage all drive, component and advanced development programmes using the DFSS methodology. The voice of the customer in the form of Critical To Quality parameters (CTQs) is assessed against existing capabilities using a flow-up and flow-down process to identify any gaps that must be bridged to provide solutions to the customers' needs.

Implementation design

The Six Sigma Academy was employed to guide the implementation and provide the initial waves of training for executives, champions and Black Belts through late 1998 and 1999. Black Belt candidates were trained in the USA and Singapore. The three-year deployment plan followed the path of manufacturing operations first, then process and support engineering, followed by design engineers, administration, sales and marketing, and then began to engage suppliers and customers as well. Over the four training phases of DMAIC each Black Belt candidate was expected to follow the 'Learn–Do' cycle, with a real project being worked on and reviewed both at the home site and in class. All sites were assigned Hands-on Champions, members of senior staff familiar with the operational requirements of the sites and trained in project selection and support.

Seagate has now developed and customized training materials so as to be self-sufficient in training up to the Master Black Belt level. Training centres of excellence exist in the USA, Europe and Asia-Pacific areas. Green Belt training is now required for all Seagate's professional and technical staff.

DFSS training was rolled out within design centres and functions, then through the advanced technology groups from 1999 to 2001.

Key benefits achieved

Go back to the year 1998 at Seagate. Upon hearing the term 'Six Sigma', the majority of employees probably stared blankly and asked, 'What in the world does that mean?' Four years later the snapshot is much different – Six Sigma, a household word at Seagate, is now highly visible and is producing impressive results for the company, both in hard savings and in improved business processes.

In hard savings, Seagate has achieved nearly $700 million. All savings are validated independently and audited by the finance team, using very strict criteria. It is also certain that the so-called 'soft savings' that Seagate has achieved but not counted far exceed this value.

Hard savings, although an essential metric to track the progress of the activity, are only a small part of the story. Seagate's operational performance has improved tremendously on an overall basis. The company has moved into a technology leadership position that has in turn led to improved market share in an ever more demanding customer environment. While Six Sigma is not the sole contributor to this impressive performance, it has been an essential enabler. Six Sigma enables top-line growth as well as bottom-line savings.

As Six Sigma matures at Seagate, its leaders are shifting their focus from reactive to proactive measures – 'We've done quite well at cleaning up problems, so now we can evaluate and prevent potential problems,' says Jeff Allen, Vice President of Six Sigma and Design For Six Sigma. This leads directly into the increasing use of DFSS, to leverage the effort earlier in the product lifecycle.

Here are some measures of Seagate's progress:

- Over 600 Black Belts have been trained or are in training, with many now reintegrated into business functions after their two-year full-time assignment.
- Over 2700 Six Sigma projects have been completed.
- Green Belt training is nearly complete, with over 5500 employees trained.
- Over 1500 Design for Six Sigma engineers have been trained or are in training.
- Seagate's Six Sigma projects have delivered nearly US$700 million of cumulative, validated financial savings.

Lessons learned

The main learning points from the Six Sigma programme at Seagate Technology include:

1. *Metrics of management.* Companies using Six Sigma need to learn how to use the metrics to manage – to make appropriate decisions on a holistic basis, avoiding sub-optimization. This task of integration with the whole of the company's business process is the key.
2. *Goals.* Set aggressive goals – don't make them too easy.
3. *Soft savings.* Develop a system for tracking 'soft savings'.
4. *Common language.* Develop a common language and encourage its use on a widespread basis early in the programme.
5. *Train all functions.* Embed the business process within the organization by training all functions – use Green, Black Belt and customized programmes as appropriate.

Company leaders are now asking: What is the next step for this programme? Are employees ready to take that step? And how do we get there?

In April 2002, a diverse group of representatives gathered for a two-day, intensive planning session to help define Seagate's future strategy for Six Sigma. The fourteen representatives formed a collective vision for Six Sigma's future over the next three years, and included specific elements that will help to build a solid basis for future growth. Some of the input included:

- Integration – Seagate is at a point where the Six Sigma effort can start shifting from being viewed as an initiative to being integrated into the way the company does business. It becomes embedded within the organization.
- Metrics – critical processes will be well understood and their performance tracked with customer-focused, variation-based metrics. These metrics will guide company strategy and behaviour well beyond current implementation, taking Seagate to the next level of managing via metrics.
- DFSS – all products being developed will use the Design for Six Sigma process, enabling time-to-market and time-to-volume consistency and predictability. The DFSS process will integrate all functional groups into a global new product-delivery team.
- Projects – a dynamic process for prioritizing projects and allocating resources will exist to verify that Six Sigma projects are addressing current critical business issues.

Future plans include an ongoing assessment of the programme to make sure Six Sigma is delivering what it is intended to deliver. 'Six Sigma is becoming embedded in our everyday activities in many areas thanks to the work of many employees who embraced this program,' President and COO of Seagate Bill Watkins said. 'Our dedication to Six Sigma can be seen in the financial returns and also in the pervasiveness of the program throughout the company'.

(Information provided by members of the Seagate Technology Six Sigma team, supplied by Rob Hardeman and Clare Desmond of Seagate Technology, Ireland.)

5

Lean enterprises

Vision without action is a daydream
Action without vision is a nightmare.

Japanese Proverb

Introduction

In this chapter the concept of lean enterprise and its impact on Six Sigma are considered. It is shown that lean enterprise began in car manufacturing in Japan, but that today a lean enterprise is any organization that has largely eliminated any activity that absorbs resource but does not add value to the product or service. The requirements for lean are considered and are married to the concepts of Six Sigma. We conclude by showing how the benefits of Six Sigma and lean can be combined with FIT SIGMA™ to keep an enterprise lean and fit.

Origin of lean

As with all facets of the quality movement, the origin of lean enterprise is in manufacturing. Lean enterprise philosophy (and make no mistake, lean is more than a system, it is a philosophy) began with Japanese automobile manufacturing in the 1960s and was popularized by Womack and co-authors in *The Machine that Changed The World* (1990), which is essentially the story of the Toyota way of manufacturing automobiles. Up until then the manufacturing of automobiles had changed very little since Henry Ford adapted the conveyor belt for manufacturing cars in 1913. Prior to Henry Ford's assembly line, the automobile had been a luxury item handmade by a group of workers in a stationary workplace. Ford's conveyor-belt (the assembly line) approach allowed production to take place on a moving belt with each worker doing a small, specialized task. Ford believed that if each step of production was broken down to the smallest element, then 'the stupidest man could become a specialist in two days'. With this moving conveyor-belt approach Ford was able to produce 250 000 cars a year, which sold at $500 each. From being a luxury item that only the rich could afford, the car now became in effect a consumer item within the reach of most families. The downside was the minute division of labour, the cyclical nature of the work, and the inexorable pace of the moving conveyor belt. Workers lost a sense of purpose of what they were doing; they could not see that they were building cars but rather saw a repetitive, mindless task, such as putting bolts on a component as it moved past them. As it says in *The Man on the Assembly Line* (Walker and Guest, 1952):

> *The assembly line is no place to work, I can tell you. There is nothing more discouraging than having a barrel beside you with 10 000 bolts in it and using them all up. Then you get another 10 000 bolts and you know that every one of those bolts has to be picked up and put in exactly the same place as the last 10 000 bolts.*

Chrysler, General Motors and other manufacturers soon adopted the assembly-line approach, but whereas Ford only had one model (the model 'T') the others, led by General Motors and Chrysler, began offering several models in the 1920s. Ford had to follow suit, and to do so had to cease production for seven months while new models were rushed into production. The assembly-line approach was still used and models were made in batches. Changing a model required set-up time for the change of dies etc. Work at each stage of production was still broken down to the lowest level, workers were not expected to think, and there was a heavy reliance on inspection and testing to maintain the standard of the finished product.

The next major change in car manufacturing is credited to Ohno Taiichi of Toyota. Ohno Taiichi returned to Japan after visiting car manufacturers in the USA in the 1960s and developed a new method of manufacturing, which became known as lean production.

The main characteristic of lean production, sometimes referred to as Toyotaism, is that materials flow 'like water' from the supplier through the production process and on to the customer with little if any stock of raw materials or components kept in warehouses, no buffer stocks of materials or part-finished goods between stages of the manufacturing process, and no output stock of finished goods. This 'just-in-time' approach requires that materials arrive from dedicated suppliers on the factory floor at the right stage of production just when required, and when the production process is completed it is shipped directly to the customer. With no spare or safety stock in the system there is no room for error. Scheduling of activities and resource has to be exact, communication with suppliers must be precise, suppliers have to be reliable and able to perform to exacting timetables, materials have to arrive on time and meet the specification, machines have to be maintained so that there is no down time, operators cannot make mistakes, there is no allowance for scrap or rework and, finally, the finished product has to be delivered on time to customers. This is often implemented by circulating cards, or *kanbans*, between a workstation and the downstream buffer. The workstation must have a card before it can start an operation. It can pick raw materials out of its upstream (or input) buffer, perform the operation, attach the card to the finished part, and put it in to the downstream (or output) buffer. The card is then circulated back upstream to signal that the next upstream workstation should carry out the subsequent cycle. The number of cards circulating determines the total buffer size. *Kanban* control ensures that parts are made only in response to a demand. With computer-controlled production, the *kanban* principle applies but there is not a physical movement of cards; information is transferred electronically.

This 'just-in-time' approach generally precludes large batch production; instead items are made in 'batches' of one. This means that operators and the system have to be flexible, and 'single minute exchange of dies' (SMED) becomes the norm. A lean approach reduces the number of supervisors and quality inspectors. The operators are trained to know the production standards required and are authorized to take corrective action – in short, they become their own inspectors/supervisors. The principles of Total Productive Maintenance (TPM) and five Ss (from a set of Japanese words for excellent housekeeping, see below) are followed, and as a result the equipment becomes more reliable and the operator develops 'ownership' towards the equipment.

Another important aspect of the Toyota approach was to expand the work done at each stage of production. For example, a team of workers is responsible for a stage of production (or 'work cell') on the moving assembly line, such as installing the transmission or the seats, etc. Each team is responsible for its part of the assembly, and it may be able to make minor changes to procedures within the confines of a time limit (the time allowed on the moving line for production to move from one stage to the next) and specified standards – for example, the team can change the order of assembly at its workstation but does not have the authority to add extra nuts, etc. Quality standards are assured with the application of zero quality control or quality at source before the actual production begins, and *poka yoke* (mistake proofing) during production.

The original Toyota model of Lean Manufacturing, from which various hybrids were developed, comprised eight tools and approaches:

1. Total Productive Maintenance (TPM) – see Basu and Wright (1998), pp. 96–99.
2. Five Ss – these represent a set of Japanese words for excellent house keeping (*sein*, sort; *seiton*, set in place; *seiso*, shine; *seiketso*, standardize; and *sitsuke*, sustain).
3. Just in Time (JIT)
4. Single Minute Exchange of Dies (SMED)
5. *Judoka* or zero quality control
6. Production work cells
7. *Kanban* (see above)
8. *Poka yoke*.

These terms, and others, are explained in the glossary at the end of the book.

Muda

A visitor to a lean manufacturer will be struck by the lack of materials; there is no warehouse, no stocks of materials between workstations, and no stocks of finished goods. At first glance this suggests that lean is an inventory system. However, lean is not just an inventory system; it also means the elimination of *muda*. *Muda* is a Japanese word that means waste, with waste being defined as any human activity that absorbs resource but creates no value, and thus the philosophy of lean is the elimination of non-value-adding activities. The rough rule is the elimination of any activity that does not add value to the final product, and the taking of action so that the non-value activity never occurs again.

Before anything can be eliminated it first has to be identified. The Toyota approach to identifying areas of waste is to classify waste into seven *mudas*:

1. Excess production
2. Waiting
3. Movement or transportation
4. Motion
5. The process
6. Inventory
7. Defects.

The approach is to identify waste, find the cause, eliminate the cause, make improvements, and standardize (until further improvements are found). The usual approach is akin to flow process charting (as used by industrial engineers) to show operations, transports, delays and storage. A flow process chart is shown in Figure 5.1. Once the process has been charted each activity is queried as to why it happens, with the aim of elimination or improvement. Questions are asked in the order what, where, when, who and how – i.e. what is done and why is it done (what would happen if it wasn't done), what else could be done and what should be done, when is it done (why that sequence), where is it done and why there, who does it and why that person, and how is it done and why that way. This approach, the basic problem-solving technique, is shown in Table 5.1. Obviously all of these questions can equally be applied to a non-manufacturing operation, and if asked in an office where there has been a bureaucratic culture the results can be quite startling. In organizations where we have consulted we have found whole departments repeating and recording information that has already been recorded elsewhere, and as if that was not bad enough we have often found that much information is collected and recorded but is never referred to (let alone adding value to a good or service!).

What is a non-value-adding activity?

If the concept of elimination of non-value-adding activity was taken literally, then what of overhead departments and support departments such as finance

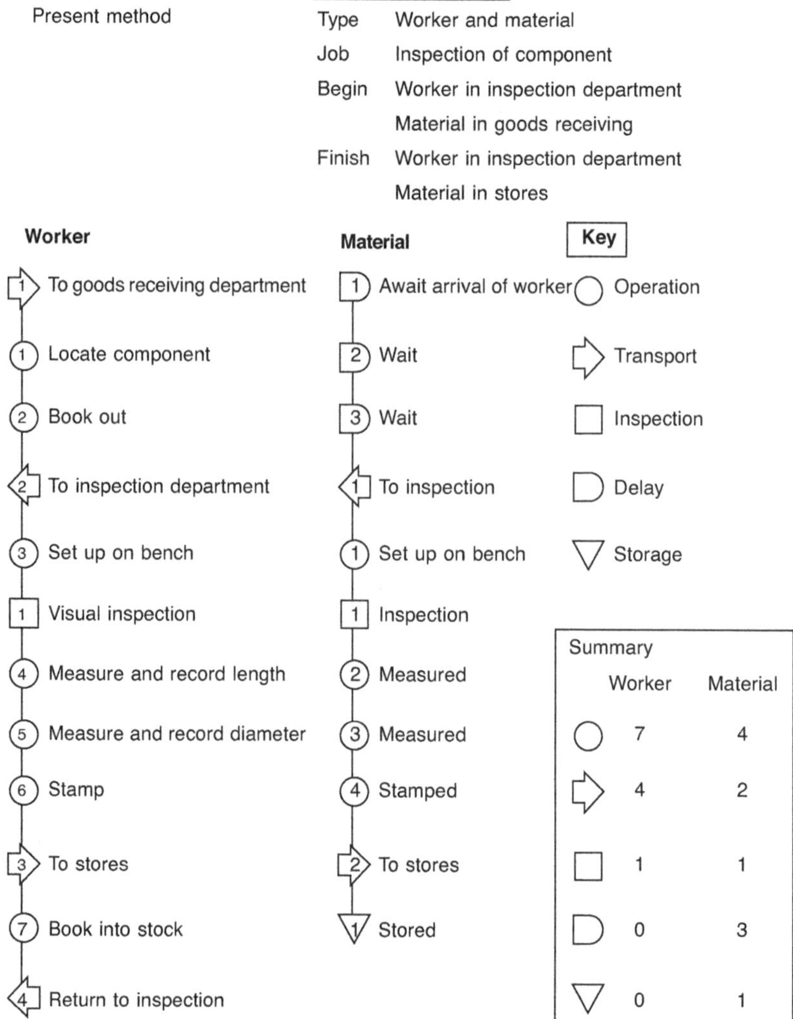

Figure 5.1 Worker and material flow process chart.

Table 5.1 Basic problem-solving chart

	Why?	
• What?	Why?	What else?
• Where?	Why?	Where else?
• When?	Why?	When else?
• Who?	Why?	Who else?
• How?	Why?	How else?

and human resources – what direct value do they add to the product or to the customers? And what value is added to the product or service by having pot plants in the staff cafeteria? Interestingly, when a company goes into receivership, often the first department to be closed by the receiver is the human resource department, which suggests that receivers and statutory managers do not see the human resource function as adding value. However, let us accept that some activities are necessary and cannot be eliminated even though they do not directly add value to the product or service. For example, compiling and filing the annual report does not add value to the product but nonetheless is legally necessary. Likewise the pot plants (and the cafeteria itself) do add to a pleasant environment and might help establish teamwork and a positive culture. Not all non-value-adding activities can (or should) be eliminated! (Save our pot plants . . .)

Extension of lean model

Today any operation that aspires to world class has to be lean, and to provide excellent service. Over the years the Toyota lean model has been adapted and extended to cover any enterprise, including pure service-type organizations where steps are taken to identify activities that do not add value to the product or to the service provided to the customer and, where possible, eliminated. For manufacturers the competitive difference between one manufacturer and another will be service, and for a service industry the only difference can be in the level of service provided. It follows that organizations that can produce the best products at competitive prices, and provide excellent service, will have an added advantage if resources are not being wasted on activities that do not add value.

In congruence with value-driven objectives to satisfy customers, as outlined above, one important extension of the lean model is the value stream. As shown in Figure 5.2, the value stream is a set of specific actions required to bring a product (goods or services) through the three critical business processes:

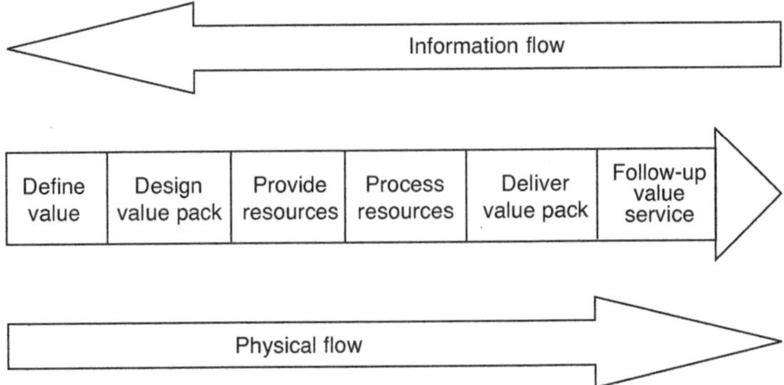

Figure 5.2 A value stream.

1. Operations – from the concept through detailed design and production to launch
2. Information flow – from order taking to detailed scheduling to delivery
3. Physical flow – from raw materials to a finished product in the hands of the customer.

On a closer analysis, the value stream contains three complementary characteristics of delivering the best value to customers. First, similar to Porter's value chain (Basu and Wright, 1998, pp.192–193), the primary activities of 'Operation' are underpinned by support activities of both 'Information flow' and 'Physical flow'. Second, similar to supply chain management principles (see Wright, 1999, pp. 45–46), departmental boundaries are ignored and the synergy of marketing, operations, accounting and human resources is applied to manage the whole process and deliver the product to the customer. Third, the model underlines the fact that defects or non-value-added activities at a later stage of the stream are more expensive than those at an earlier stage.

Characteristics of lean production

Since the 1960s, lean production has evolved and developed the following features:

- Assembly line production, but there are longer periods at each stage of production so teams of workers are now involved in installing a complete component – i.e. fitting the seats into a car and testing them – rather than an individual tightening a single set of nuts.
- Integrated production from the supplier through the manufacturing process to the customer.
- Suppliers treated as members of the production team. Suppliers are appointed on the basis of reliability, delivery on time and to specification, rather than on price alone. This extends back down the supply chain to the suppliers of suppliers.
- Just-in-time delivery of materials from suppliers. This results in the elimination of warehouses of input materials, and buffer stocks of materials between stages of production.
- Finished goods are not held, but are delivered direct to the customer.

The emphasis is on prevention of mistakes rather than detection of errors by inspection. Prevention is achieved by workers knowing the standards and self-testing, rather than relying on independent checkers. The principles of lean are equally applicable to service industries.

Lean for non-manufacturing operations

For non-manufacturing operations, lean means the elimination of any wasteful activity (*muda*) that does not add value to the service provided to the customer. Typical examples of *muda* in a service operation such as secretarial administration are:

- Excess production – preparing reports that are not acted upon
- Waiting – processing monthly or in batches, and not continuously
- Transportation – fax machines and printers are at a distance from the workstation
- Motion – steps and data entry
- The process – signoffs
- Inventory – transactions not processed, work waiting to be done
- Defects – incorrect data entry/typing mistakes/misfiling.

Staff are encouraged to take responsibility for their activities and be proactive, and to make suggestions for improvement of process (more efficient use of time and resources) and enhancement of the level of service provided to the customer. Teamwork, open communication and flexibility are accepted as part of the culture.

What does the customer want?

The focus of lean is on providing an improved product or better service to customers. To achieve this objective, the organization has to know what the customer wants and what the customer values in terms of specification, cost and timing. Knowing what the customer wants is one aspect, but knowing how well we perform to the customer's requirements is another issue. Assuming that we have made an effort to find out exactly what the customer values, then the issue is measuring our performance. Thus some key performance indicators have to be identified, and a method of monitoring performance devised. Obviously this is compatible with Sigma. With Sigma, areas to be measured are identified, standards are set, and the aim is to reach Six Sigma in performance. Where Six Sigma is not achieved, actions are taken to find the causes of non-performance with the aim of elimination of the cause.

Example 5.1 American Aluminium Industry

Author and researcher Dr James P. Womack, president of the Lean Enterprise Institute, told attendees at a meeting of the Aluminium Association that companies that produce aluminium are among the many basic industries that could do a lot more with less difficulty and lower costs if they did some lean thinking.

Womack described the labyrinth and costly processes associated with delivering an aircraft, automobile or drinks can, and suggested that manufacturers think about the consumer first and work backward to gain efficiencies and cut out the non-value-adding activities:

Ask, from the customer's standpoint, what is of value among your activities? What is wasted? How can we eliminate the waste? It's so simple and yet so very hard to do. Most people are in love with their assets, technologies and organization.

He cited a military aircraft programme as an example:

The typical subassembly goes through four plants, four states and 74 organizational handoffs between engineering, purchasing and fabricating operations. It goes 7600 miles and takes darn near forever. It takes two to four years from beginning to flyaway condition.

Womack pointed out that manufacturers need to identify the value stream for each of their products, and document all the steps it takes to get from raw material to the customer:

Step 1, get the value right. Your customer is not interested in your assets. He is interested in his value. Step 2, identify the value stream from start to finish, not just within the walls of your plant or company.

He advised that the remaining value-creating steps should be organized so they impact on the product in a continuous flow:

If, and only if, you can create flow, then you can move to a world of pull. You put the forecast in your shredder and get on with your life. And you make people what they want. The customer says, 'I want a green one'. And you say, 'Here's a green one'. That's a very different world from your world of endless forecasts, always wrong, and the desperate desire to keep running, which makes you produce even more of the wrong thing because that makes the numbers look good in the short term.

(Case reported in *Metal Center News*, May 1999, Vol. 39(6), p. 123.)

Lean and Six Sigma

Monitoring performance can, in most areas, be straightforward in the factory. Within a factory it is easy to measure if production is to specification, if finished goods are delivered on time, and if they are delivered in the right quantities. Such measurement should begin with the receipt of input materials, and suppliers should be judged on performance to Six Sigma. Six Sigma measures can be set for suppliers of inward goods through every stage of interaction, including the final stage of payment – i.e. are invoices correct?

In the production process itself, stocks of raw materials, waste and scrap in production, idle operator time, buffer stocks of materials between processes, down time of machines, the production cycle time and the costs of each failure to conform to the standard are all areas that lend themselves to specific Six Sigma projects.

Once the goods have been produced, the next stage is shipment to customers. Again there are obvious easy measures here – e.g. the goods shipped meet the specification, are delivered on time, with the right quantity, and invoicing is accurate. All these activities are easily identified, and performance to Six Sigma should be simple to measure. Other measures can be set after consultation with customers; it is important that we know what the customers value and that we measure our performance against agreed customer criteria.

Knowing what to measure is just the start; recording performance requires a process and some discipline. Measurement alone is not sufficient unless some action is taken. Measurement is taken to show where performance is below the Six Sigma standard. Once a variance has been found, then action has to be taken to find and eliminate the causes. With lean organizations there is no room for errors, and the Six Sigma project approach is an obvious method of eliminating them. Lean goes further than basic Six Sigma, as it requires an understanding of where value is being added and which activities are non-value adding.

Working with suppliers

With Lean production the aim is to achieve a just-in-time system. 'Just in time' means that materials are received directly into production from suppliers just when required. This means that suppliers have to be geared up to deliver to the right specification and on time. With just-in-time manufacturing there is no room for errors in specification, or for late delivery. All of this requires the closest cooperation and teamwork with suppliers. A truly lean organization manages the supply chain right back to the original equipment manufacturer – in short suppliers become part of the team and, on top of the obvious measures of meeting specification and delivery times, are valued for loyalty, proactive advice and service. Indeed, key suppliers often become members of the design team for new products/services. Specification and timing can be measured using Six Sigma criteria, but attitude and cooperation is not so easy to measure and is often overlooked in the mechanistic approach of Six Sigma.

Although with lean cost is not the key measure, nonetheless cost is still an important issue. Thus with a lean organization, cooperation with suppliers means the lean organization lending Six Sigma trained staff to bring the supplier up to Six Sigma status, or to use the Six Sigma project approach to identify and eliminate non-value-adding activities within the supply organization. Toyota and McDonalds, although not Six Sigma organizations, have a long tradition of 'lending' their quality and efficiency experts to key suppliers to improve performance and lower cost. With McDonalds the emphasis is on consistency of product – indeed at one stage in Europe the potatoes were not up to standard, and McDonalds resorted to airfreighting potatoes from Canada until consistency could be achieved from European-grown potatoes. In this case the McDonald scientists worked closely with the European growers to achieve a potato of the required standard. Toyota is not only concerned with the consistent quality of parts (this is now taken for granted by the suppliers and Toyota) but also with getting suppliers' costs down so as at least to hold, if not reduce, the price of parts. Toyota are more than willing to 'lend' teams of experts to show the supplier how to perform better.

Lean Sigma and FIT SIGMA

As indicated in Chapter 1 and discussed further in this chapter, the industrial engineering tools and Toyota approach of lean enterprise complement the rigour and statistical processes of Six Sigma. It is the integration of these tools and this approach that addresses the entire value delivery system known as Lean Sigma, or Lean Six Sigma. In a simplified model (see Figure 5.3) the variation control of Six Sigma added to the waste control of lean enterprise has led to Lean Sigma, and when sustainability is added we get FIT SIGMA.

Example 5.2 Lean Sigma in GlaxoWellcome

Before the merger with SmithKline Beecham to form GSK, the pharmaceutical multinational GlaxoWellcome set out to establish Manufacturing and Supply as a core competence in 1999, and a major driver to achieve that was Lean Sigma. Drawing from the learnings of Six Sigma programmes and Lean Manufacturing successes, the company developed a worldwide programme to roll out Lean Sigma as a transformational and continuous improvement process for the Manufacturing and Supply Directorate comprising over 19 000 employees.

The declared principles of Lean Sigma were:

1. To specify *value* in the eyes of the customer
2. To identify *value stream*; eliminate *waste and variation*
3. To make value *flow* by pull from the customer
4. To involve, align and *empower* employees
5. *Continuously to improve knowledge* in pursuit of perfection.

The core content of the programme was the training of 175 Experts (equivalent Black Belts) supplemented by Leadership training. The Experts had target cost savings projects and the Leadership themselves had the responsibility to lead 'Just Do It' projects. To bolster these continuous improvement efforts, the training of 900 Advocates (equivalent Green Belts) was planned. The training and consultancy support was provided by a group of consultants led by Air Academy from Colorado, USA.

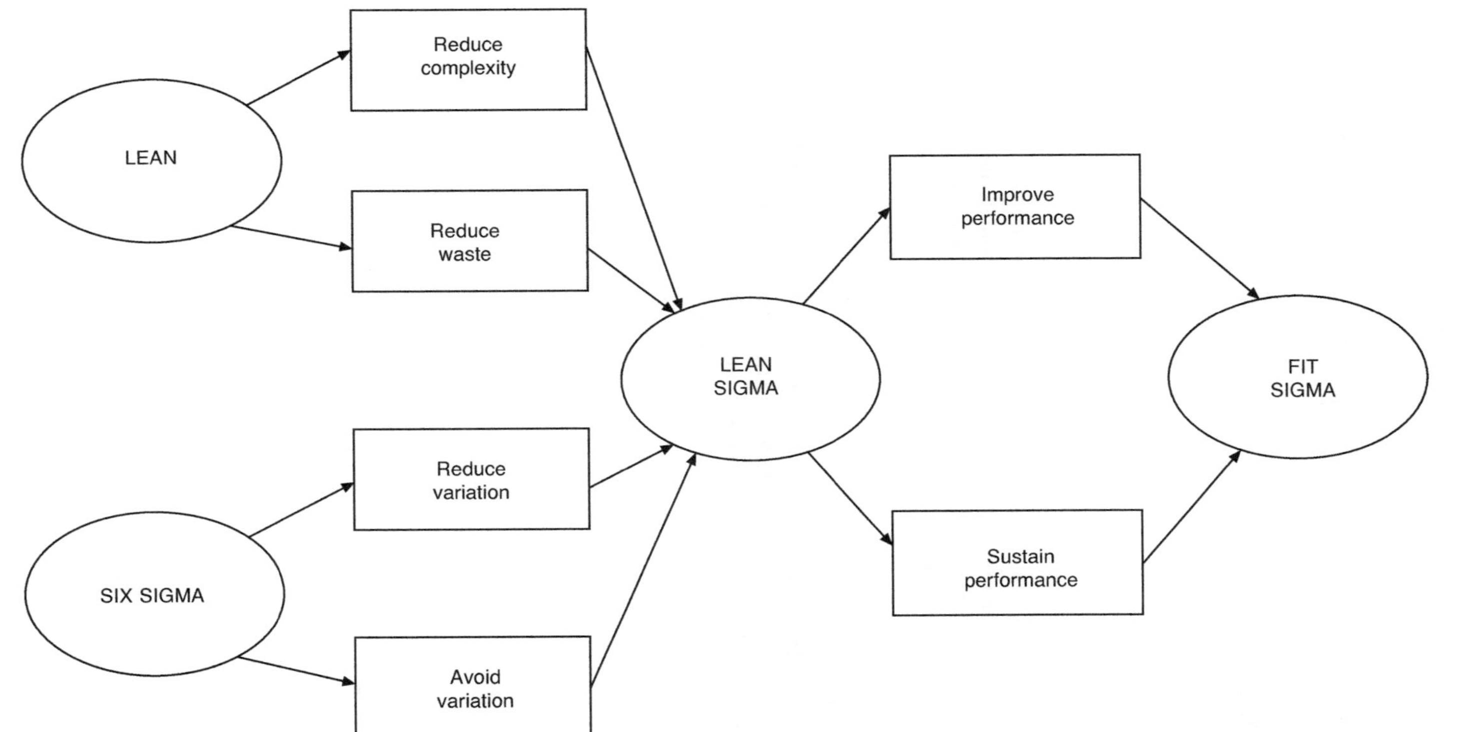

Figure 5.3 Quality beyond Six Sigma.

> Arguably, GlaxoWellcome (GW) was one of the first users of the integrated Lean Sigma approach. Although the programme was not in place within GSK when the merger took place in 2001, the fact the GW programme was retained by GSK is a testament of success.

Summary

The outward and visible result of lean in manufacturing is the absence of inventory, but this is only an indication of what lean actually means. Lean means the identification and elimination of non-value-adding activities. Lean has a customer focus, but also includes dedicated suppliers, and the closest of co-operation with suppliers.

Sigma, when combined with lean, includes measurement of performance through the production process (from the receipt of materials through to the finished product) to Six Sigma performance. Six Sigma project teams are used to identify problem areas and to eliminate non-value-adding activities. Lean enterprises include not only manufacturing organizations, but also any organization that has the philosophy of identifying and eliminating non-value-adding activities.

It is not surprising that organizations, whether they are from the manufacturing or service sectors, are moving more and more towards Lean Sigma rather than pure play Six Sigma. This hybrid provides both robustness and flexibility in the programme, and thus is a good foundation for FIT SIGMA.

6

The methodology of FIT SIGMA™

Suit the action to the word, the word to the action.

Shakespeare, Hamlet Act 3

Introduction

The success of Six Sigma and Lean Sigma cannot be faulted. The rigorous Six Sigma process combined with the speed and agility of Lean Sigma has produced definitive solutions for better, faster and cheaper business processes. Through the systematic identification and eradication of non-value-added activities, an optimum value flow is achieved, cycle times are reduced and defects are eliminated, with the result of an improved all important bottom line.

However, managers confused by the grey areas of distinction between different quality initiatives and challenged by increasing internal demands and external competition are expressing concerns, including the question 'how do we sustain the results?' This chapter answers this question and shows how to maintain and build on initial benefits. We call this 'sustaining fitness'.

Barriers to achieving and sustaining results

A recent survey by Basu (unpublished GSK survey, January 2001) has shown that there are considerable barriers to achieving and sustaining results in quality initiatives. These are illustrated in Figure 6.1.

The biggest obstacle appears to be the packaged approach of quality programmes, causing a paucity of customized local solutions. Furthermore, due to the 'top-down' directive middle managers are often not 'on board' – the initiative is not owned by employees. We can identify additional and complementary areas of concern, including:

- 'Some star performers of Six Sigma have shown poor business results' (e.g. site closures by Motorola)

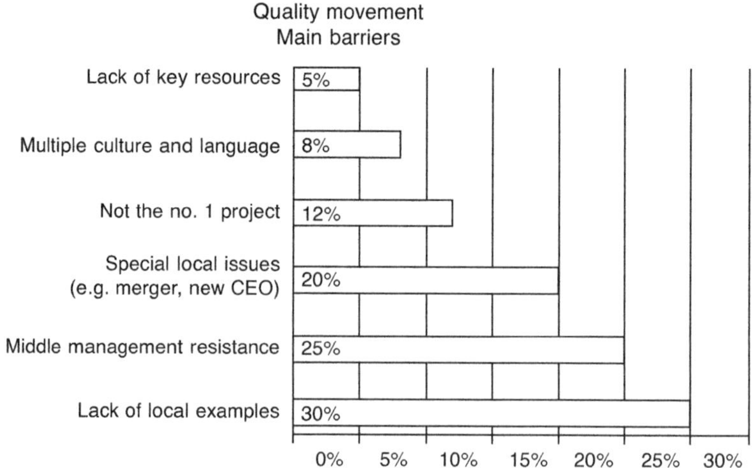

Figure 6.1 Main barriers to quality movement.

- 'Incomplete initiatives' (e.g. Marconi abandoned Six Sigma during the economic downturn of 2001)
- 'Change of management and loss of sponsors' (e.g. the decline of Six Sigma at Allied Signal after the departure of Larry Bossidy)
- 'External push by high-powered consultants' (e.g. the dominance of the Air Academy consortium in the GSK programme)
- 'Excellent early results not sustained' (e.g. Ratheon relaunched Lean Sigma after a drop in performance)
- 'High start-up costs impede small- and medium-sized enterprises' (the initial training start-up cost for Six Sigma is reported to be over $1 million)
- 'Still regarded as tools for manufacturing' (in spite of the success of Six Sigma in GE Capital).

The dramatic bottom-line results and extensive training deployment of Six Sigma and Lean Sigma must be sustained with additional features for securing the long-term competitive advantage of a company. If Lean Sigma provides agility and efficiency, then measures must be in place to develop a sustainable fitness. The process to do just that is FIT SIGMA. In addition, the control of variation from the mean in the Six Sigma process (σ) is transformed to company-wide integration in the FIT SIGMA process (FIT Σ). Furthermore, the philosophy of FIT Σ should ensure that it is indeed fit for purpose for all organizations – whether large or small, manufacturing or service.

FIT Σ is a solution for sustainable excellence in all operations. It is a quality process beyond Six Sigma. The fundamentals of FIT Σ are underpinned by three cornerstones, as shown in Figure 6.2.

Taken individually, the main components of these cornerstones are not new, but they do constitute proven processes. What is new is the combining

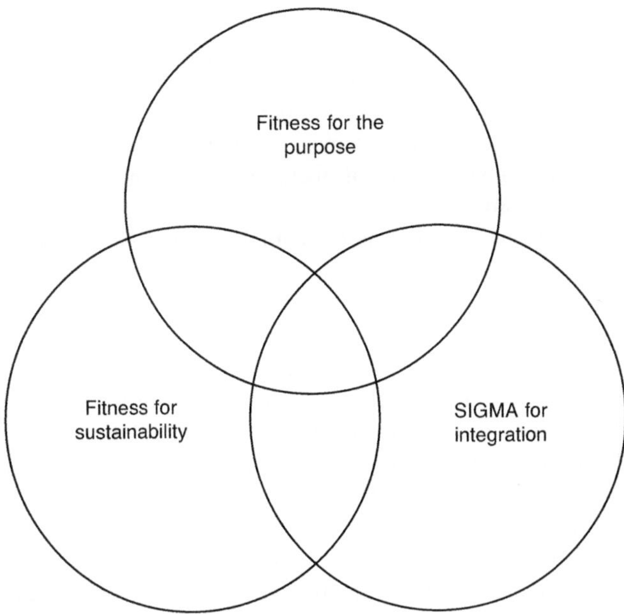

Figure 6.2 Fit Σ fundamentals.

of these components, and thus the total FIT Σ is both novel and unique. The elements are:

1. Fitness for the purpose
   - Initial assessment
   - For all functions
   - For any size of organization.
2. Sigma (Σ) for improvement and integration
   - Appropriate Six Sigma tools
   - Learning deployment
   - Project plan and delivery
   - Shift from variation (σ) to integration (Σ).
3. Fitness for sustainability
   - Performance management
   - Senior management review (S&OP)
   - Self-assessment and certification
   - Knowledge management.

Fitness for the purpose

It was Joseph Juran who coined the phrase 'fitness for the purpose' (Juran, 1989) relating to the basic requirements for quality. In the context of FIT Σ, 'fitness for the purpose' has wider implications. Here, fitness means that the FIT Σ methodology is tailored to fit all types of operations (whether manufacturing, service or transport) as well as all sizes of organizations (whether a multi-billion dollar global operation or a small local enterprise). The customization of the improvement programme to identify the right fit appropriate to the type and size of operation is determined by a formalized initial assessment process.

It is advantageous, although not essential, to apply the initial assessment process based on a set of questions that could be applied subsequently for the periodic self-assessment or certification process. The certification process, which will be described in more detail later, may be adapted either from international quality awards (such as the European Foundation of Quality Management, or America's Malcolm Baldridge Quality Award), a holistic, published checklist (such as Basu and Wright's Total Solutions 200 Questions; Basu and Wright, 1998), or the organization's own checklist. Regardless of which checklist is adopted, it is essential that it is customized and that the key players of the company believe in it.

As part of the Six Sigma or Lean Sigma programme, a 'baseline analysis' is carried out to identify areas of improvement after committing the company to a rigorous deployment programme. For FIT Σ the initial assessment is similar to that of a baseline analysis, but it is carried out right at the onset *before* the start of the deployment plan.

We believe that there are substantial benefits to be gained from the organization designing its own assessment process rather than following a consulting firm's standards. However, it is imperative that the company is aware of the requirements and criteria of the assessment. The initial evaluation requires a good understanding of the company's own processes, with the objective of training to international standards. It is necessary that the assessor is trained to know what to measure and what international (world class) standards are.

Example 6.1 Six Sigma in a small and medium enterprise

The Solectron factory in Ostersund, Sweden, where AXE switchboards are manufactured, employs approximately 1000 people. The site was formerly part of the Ericsson Network of core products. Solectron, as an independent company, was experiencing tough competition even at the crest of the 'telecom boom'. With the downturn of the market from 2000, the competition became increasingly fierce. The management were toying with the idea of launching a Six Sigma initiative, but their initial enquiry revealed that they would be set back by at least $1 million if they began a formal Black Belt training programme with the Six Sigma Academy at Scottsdale, Arizona.

Ericsson, the parent company of previous years, had already embarked upon a Six Sigma initiative. The Black Belt training programme was also in full swing. Solectron decided to send a promising manager to a

Black Belt training course via the Ericsson deployment plan. The young manager duly returned to Ostersund with great enthusiasm and applied a preliminary 'base line analysis' rooted in a simple checklist. The results were then presented to the management, and a customized programme was drafted. The training programme was extensive, but Solectron relied on the Black Belts from Ericsson and also retained the same consultants as and when required.

Ten members of the top management team attended a one-day course on Six Sigma, fourteen people were trained as 'Black Belts' on a seven-month part-time programme, and twenty more attended a two-day course. Six Sigma applications at this factory saved US$0.5 million during the first year of the project. This amounted to about $500 per employee, but was actually closer to a huge $36 000 per employee trained in Six Sigma methods. A modest start in terms of savings per Black Belt perhaps, but the investment was also a fraction of a 'pure play' Six Sigma initiative. More significantly, this customized approach enabled Solectron to have a launch pad to gain a much-needed competitive advantage.

Solectron applied the FIT Σ methodology 'Fitness for the purpose', albeit not consciously under the FIT Σ label.

Look first before diving!

Many organizations have dived into a Six Sigma programme without going through the earlier crucial stages of identifying the real requirements. Arguably, a significant number of companies that initiated a Six Sigma programme did so because they felt threatened in terms of their very survival, or they became victims of management fads. In some cases Six Sigma was attempted as a panic attempt when it was too late, and disappointment was expressed when the bottom-line and share price did not miraculously improve within a short period.

The GE factor

For many, the well-publicized successes of General Electric with Six Sigma – 'the GE factor' – was too strong to ignore. The 'GE factor' should not be interpreted as a licence blindly to copy General Electric, but GE should be seen as an exhortation to 'get effective'. This is best achieved by learning from General Electric, applying the FIT Σ methodology of what is fit for you, and understanding how to stay fit.

Starting point

The starting point is an initial assessment, as illustrated in Figure 6.3.

The initial assessment process can benefit from the application of a rating scale of 1 to 5 for each criterion or question, with 1 being 'poor' and 5

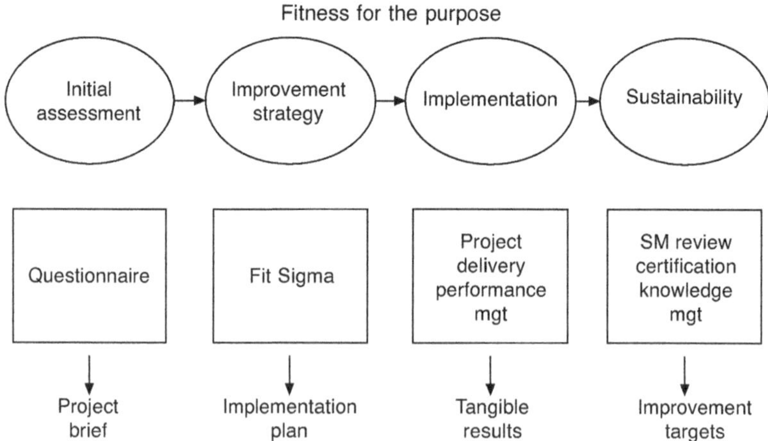

Figure 6.3 Fitness for the purpose.

'excellent'. A spider diagram can be constructed from the scores to highlight the gaps. The assessment process is described in more detail later in this chapter (see also Figure 6.14).

The 'fitness for the purpose' philosophy also applies to the type of business, whether it is manufacturing operations in a factory or a service procedure in an office. The success of GE Capital after the application of Six Sigma is well documented. Jack Welch has, however, often been quoted as the success factor rather than the application of the principles of Six Sigma; additionally, it has to be remembered that GE Capital is a huge enterprise. That FIT Σ methodology is applicable to a relatively small service operation is shown in the following example of a legal practice.

Example 6.2 Six Sigma in a legal practice

Countrywide Property Lawyers (CPL) is a subsidiary of the Countrywide Assured Group in the UK, and was established in 1997 to provide a specialized conveyancing service on residential property. Five CPL centres were opened in Woking, Northampton, Brentwood, Cardiff and Manchester. CPL has fewer than 400 employees, but it is the largest conveyancing company in the UK.

Conveyancing is fast becoming a commodity product that can be bought 'off the shelf'. Flexibility, the use of IT enablers and quality customer service are the differentiating factors in a competitive market worth £400 million in England and Wales.

Part of the current conveyancing process is examination of the quality of the data and information, including the review of contracts, title deeds and financial information. At present, CPL does not know if its conveyancing process is under control. Client satisfaction questionnaires completed by customers show a large variation in conveyancing lead time of between 8 and 15 weeks. Although CPL currently reports that on average it takes 10 weeks to see a transaction through from 'instruction' to 'completion', this figure is proving to be misleading due to the large variation in the process.

CPL considered the application of the Six Sigma concept to reduce variation in lead time. The initial assessment indicated that more than 50 per cent of transactions take longer than 10 weeks and more than 15 per cent take over 14 weeks.

This is an example of where the FIT Σ methodology can be applied. The focus is to train five experts (on one each site) specifically on the connection between averages and span in variation. These experts will train other employees, and thus variation control will be built into the conveyancing process by employing the FIT Σ methodology. The target for 'span' will change from 8–15 weeks to 9–10 weeks. Losses will be minimized and customers able to make, for example, firm arrangements for a removal van.

This is not 'rocket science', nor is it intensive Black Belt training loaded with complex statistics; rather, it is organized common sense. This is FIT Σ – 'fitness for the purpose'.

Sigma (Σ) for improvement and integration

The key cornerstone of FIT Σ methodology is 'Sigma (Σ) for improvement and integration'. The 'improvement' aspect is essentially based upon the proven tools and processes of Six Sigma/Lean Sigma initiatives. While the contents of Six Sigma or Lean Sigma approaches vary according to the company, consultant or author, the common features with FIT Σ are:

- Appropriate Six Sigma tools
- Learning deployment
- Project plan and delivery
- Shift from variation (σ) to integration (Σ).

Appropriate Six Sigma tools

It is true that Six Sigma or Lean Sigma tools are not original. For example, the focus on variation is historically known as the control chart of Deming (1982) and Shewhart (1931). Design of experiments can be linked to Taguchi's methods. Likewise, the proactive use of Pareto's 80/20 analysis and Ishikawa's 'fishbone' diagram in Six Sigma is laudable, but these are scarcely new. The flow process chart of Lean Sigma is also a classic industrial engineering tool. We also do not propose to introduce any so-called 'new' 'FIT Σ tools', but rather refer to them as 'Six Sigma tools'. Thomas Edison once said, 'Your idea has to be original only in its adaptation to the problem you are currently working on'. The adaptation of the existing tools is 'appropriate Six Sigma' tools, and these include the following.

1. Basic tools:
   - Pareto analysis
   - Flow process chart
   - UCL/LCL control chart
   - Cause and effect diagram
   - Input–process–output (IPO) diagram
   - Brainstorming
   - Scatter diagram
   - Histogram
   - The seven wastes
   - The five Ss
2. Advanced tools:
   - Failure mode and effect analysis/(FMEA)
   - Design of experiments (DOE)
   - Design for Six Sigma (DFSS).

Some of the tools have been described in previous chapters, and further details of all the above can be found in the Glossary. We recommend that the learning deployment of FIT Σ should ensure both the understanding and the application of these well-developed and proven 'appropriate tools'.

Learning deployment

In order to use tools 'appropriate for Six Sigma' to achieve longer-term benefits, it is essential that an extensive learning deployment programme is dedicated to the education and training of employees at all levels. The learning deployment plan should be formulated after the 'initial assessment', and details will vary according to the 'quality level' and size of the organization. We recommend that the proven paths of previous Six Sigma and Lean Sigma programmes should be treated with respect and a deployment plan built around the outline shown in Table 6.1.

Table 6.1 Deployment plan outline

Programme	Target audience	Duration	Approximate number
Leadership education	Senior management	2 days	3–5% of employees
Expert training (Black Belt)	Senior and middle management	4–6 weeks (in waves over 6 months)	1% of employees
Advocate training (Green Belt)	Supervisors and functional staff	1 week	10% of employees
Appreciation and cultural education	All employees	2 × half days	All

Through a rigorous training deployment programme, Six Sigma contributes to the creation of a people infrastructure within an organization to roll out a comprehensive programme. The issue is not who should be trained, but rather who should be the trainer? The original source of Six Sigma training was the Six Sigma Academy, founded in 1994 in Scottsdale, Arizona. It is run by former Motorola experts Michael Harry and Richard Schroeder, and their fees have been reported to start at $1 million per corporate client. Although there are many capable consultancy firms offering training, training costs are still running on an average threshold of $40 000 per 'Black Belt' (unpublished GSK survy, January 2001). The initial start-up and training costs have prevented many small and medium-sized companies from embracing a Six Sigma programme.

It is essential that high quality input is provided to the training programme. This is usually available from specialist external consultants. At the same time, a 'turn key' consultancy support is not only expensive but also contains the risk that 'when consultants leave, expertise leaves as well'. In the FIT Σ learning deployment programme, we recommend two options:

Option 1:

- Retain consultants for, say, 3 months
- Run part of leadership education and Expert training programme
- Develop, with the assistance of the consultants, the company's own trainers and experts to complete the remaining waves of leadership and Expert training
- Train Advocates (Green Belts) using own experts
- Cultural education by line managers.

Option 2:

- Deploy consultants for one top-level leadership education workshop
- Train a small team of Experts (two to five people) as trainers, and develop a deployment plan with the assistance of consultants
- Roll out the deployment plan with the company's own experts
- Ensure that the consultants are available if required
- Train Advocates (Green Belts) using own experts
- Cultural education by line managers.

In General Electric, the cathedral of Six Sigma, the Six Sigma programme has been supported globally by a corporate team (known as CLOE – Centre of Learning and Operational Excellence), based at Stanford, Connecticut. Regardless of the type or size of operations, the development of own training capability is the foundation of sustainable performance.

Example 6.3 Six Sigma training deployment in Noranda Inc.
Noranda Inc. is a leading international mining and metals company for copper, zinc, magnesium, aluminium and the recycling of metal, with headquarters in Toronto. The company employs 17 000 people around the world, and its annual turnover in 2000 was $6.5 billion.

In August 1999, the Board of Noranda decided to embark upon a global Six Sigma project with an initial savings target of $100 million in 2000. There were some specific challenges to overcome. The company business is in a traditional industry with long-serving employees. Furthermore, Noranda is a 'decentralized' company with multiple cultures and languages. Senior executives studied the experiences of other companies (GE, Allied Signal, Dupont, Bombardier and Alcoa) and invited the Six Sigma Academy from Arizona to launch the training deployment programme.

The Six Sigma structure at Noranda focused on the training of the following levels:

- Deployment and Project Champions
- Master Black Belts
- Black Belts
- Business analysts and validates
- Process owners
- Green Belts.

The Six Sigma Academy was intensely involved for the first three months of the programme, and then Noranda started its own education and training. The training accomplishments in 2000 were impressive:

- All 84 top executives followed a two-day workshop
- 90 Black Belts were certified
- 31 Champions were trained
- There were 17 days of Master Black Belt training
- There were 3000 days of Green Belt training
- There were more than 3500 days of training in 2000 – and this has continued.

The learning deployment to train and develop your own Experts or Black Belts provides a successful balance between a well-measured job structuring by a central team of industrial engineers and self-managed work teams. Over the years the principles of industrial engineering and Taylorism became corrupted to the stage where time and motion study people found the best method and then imposed that method upon the worker. The external control impeded teamwork and created the tedium of repetitive operations. This was followed by the Quality Circles (mainly in Japan) and self-managed teams (mainly in Scandinavia). The failure of the experimental Volvo factory at Uddevalla in the early 1990s was a wake-up call to realize that the planners and team leaders should be trained in analytical tools. The experience of both GM–Toyota joint ventures in California and the Ford–VW factory in Portugal demonstrated that group performance can be improved by training the teams in industrial engineering principles. The collaboration of Expert (or Black Belt) training and the team comprising Advocates (or Green Belts) provides a balance of empowerment, motivation and measured efficiency.

Project plan and delivery

It is commonly agreed that the success of any project requires commitment by the management, planning, resources and formal reviews. A FIT Σ programme is in effect a project, and the basic rules of project management apply. The process logic of a FIT Σ programme is shown in Figure 6.4, where the positions of a project plan and delivery have been highlighted.

The structure of a project organization varies according to the operations and culture, but it must comprise some essential requirements:

- A sponsor or 'torchbearer' at the highest level of the organization
- Multifunctional project team leaders who are 'Black Belt' trained
- Two-way communication – both top down and bottom up
- RACI (Responsibility, Accountability, Consulting and Information) roles must be defined clearly.

Figure 6.5 shows an example of a typical project organization for a FIT Σ programme.

We recommend that at an early stage, following the initial assessment, a project brief or project charter is prepared to define clearly:

- Project organization
- Time plan
- Learning deployment
- Project selection criteria
- Key deliverables and benefits.

The project selection criteria cover two broad categories of projects within the FIT Σ programme; large projects (managed by Black Belts) and small 'just do it' projects.

Project selection can rely on both the top-down and the bottom-up approach. The top-down approach usually relates to a large 'Black Belt' project, and considers a company's major business issues and performance objectives. Teams identify processes, critical to quality characteristics, process base lining, and opportunities for improvement. This approach has the advantage of aligning FIT Σ projects with strategic and corporate objectives. The bottom-up approach can apply to both large and 'just do it' projects.

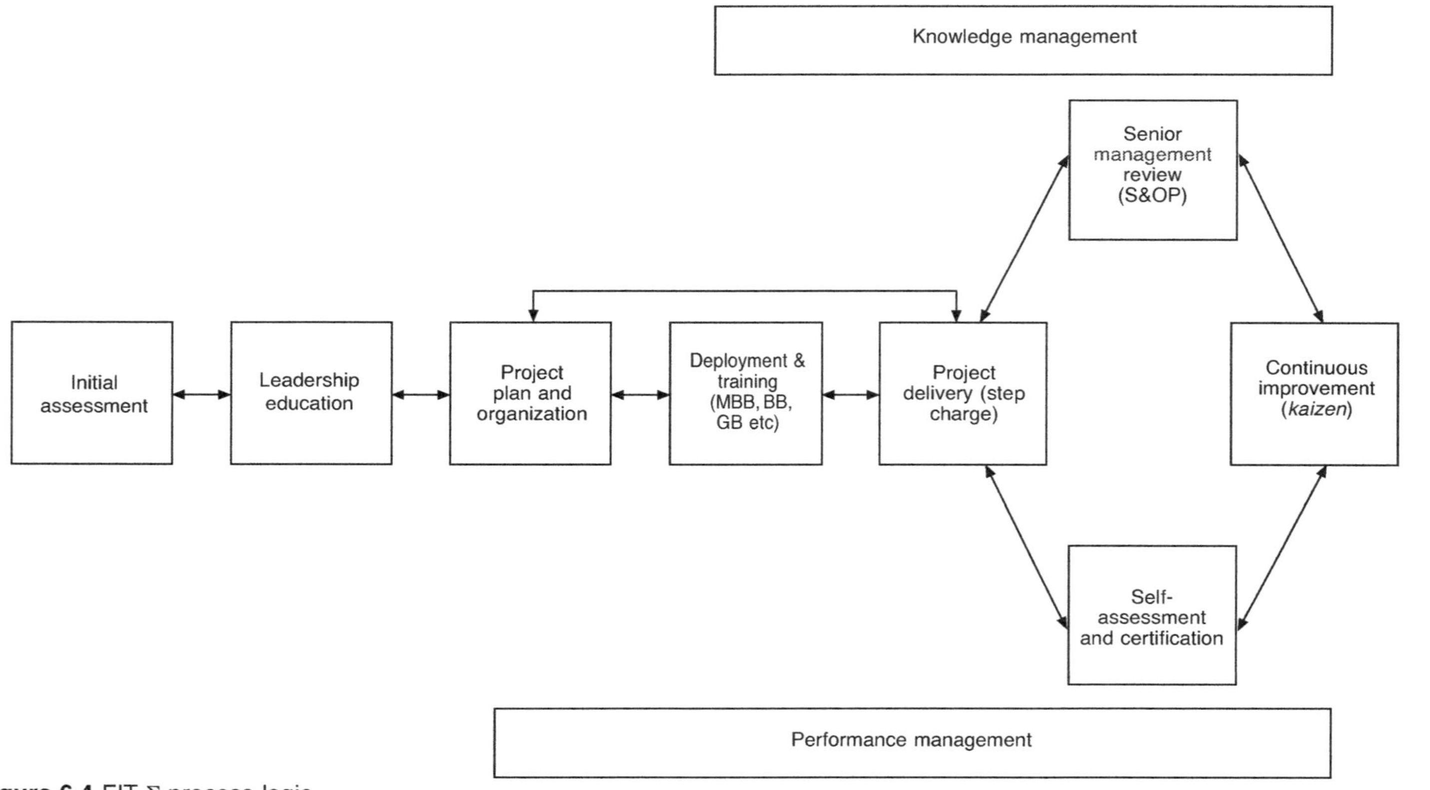

Figure 6.4 FIT Σ process logic.

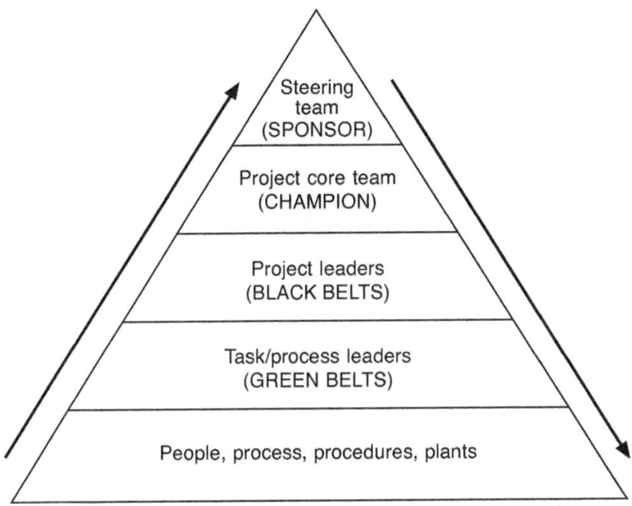

Figure 6.5 Project organization: a two-way communication process.

A rule of thumb for large Six Sigma projects is to attain savings of $1million per year per Black Belt from between four and six projects. A 'just do it' project does not usually incur a significant investment, and produces savings varying from $5000 to $100 000 annually.

Adapting from Oakland (2000), we summarized five key sources for identifying savings projects:

1. DPMO (Defects per Million Opportunities) – the number of defects per unit divided by the number of opportunities for defects multiplied by 1 000 000. This number can be converted into Sigma Value.
2. COGS (Cost of Goods Sold) – the variance is fixed and variable costs are identified.
3. COPQ (Cost of Poor Quality) – this comprising internal failure costs, external failure costs, appraisal costs, inspecting and checking costs and lost opportunity costs.
4. Capacity (overall equipment effectiveness) – the number of good units a process is able to produce in a given period of time.
5. Cycle time – the length of time it takes to produce a unit of product or service.

It is important that, at the early stage of the programme, both large and 'just do it' projects are straightforward and manageable. These types of projects are often referred to as 'low hanging fruits' (easily harvested) where quick improvements can be achieved by use of 'basic tools' (e.g. fishbone diagram, flow process charts, histograms etc.).

For example, a simple selection process has been developed by Johnson and Johnson at their Wüppertal plan in Germany. A rating of projection selection criteria related to savings and ease of implementation was applied, and the projects were selected after plotting the scores on a graph (see Figure 6.6).

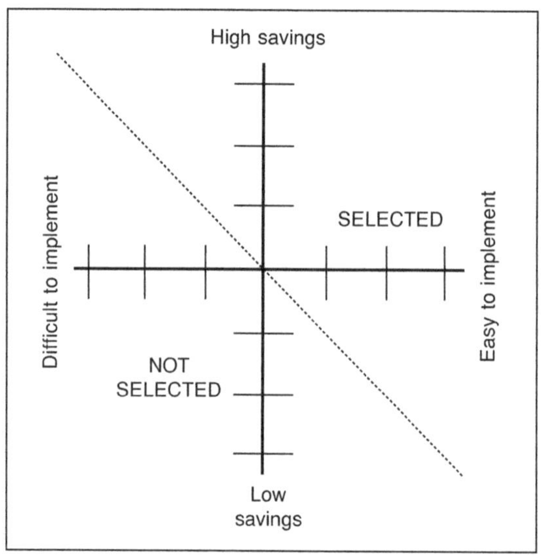

Figure 6.6 Project selection.

FIT Σ; more than a savings project

FIT Σ is more than the completion of a cost saving project. FIT Σ projects are selected and implemented to strengthen the company's knowledge base, stabilize processes and procedures, and break down cross-functional barriers.

Example 6.4 A Six Sigma project in the Dow Chemical Company
Film Tec Corp. is a subsidiary of the Dow Chemical Company in Minneapolis, and a manufacturer of water purification membranes. The quality of a membrane is determined by two criteria: flux (or the rate of water that the membrane lets through) and how much impurity is removed from the water. At Film Tec, membrane elements were tested prior to shipping to customers. The quality of membranes the customers received was protected but the speed with which they were serviced suffered. Rejected products were costing Film Tec approximately \$500 000 a year.

With a focus on customer needs, the Six Sigma project team required a strategic shift in participation by all employees to analyse systematically the internal manufacturing procedures. A key variable identified for improvement was the inconsistency in the concentration of a chemical component used in the manufacturing process. The problem stemmed from the interruption to batch feeding the chemical into the manufacturing process. To reduce the variation, an inexpensive reservoir was added to feed the chemical while containers were exchanged. Additionally, a level transmitter with an alarm was installed to alert the operators to the low level of the chemical.

The improvements have been significant. The reject rate has reduced from 14.5 per cent to 2.2 per cent. In addition to these savings for Film Tec, membranes are available to customers faster than before.

Shift from variation (σ) to integration (Σ)

Our FIT Σ process fully accepts the importance of variation reduction. The risk of an improvement process based upon average values alone has been incontrovertibly proven. Likewise, there is an abundance of real-life examples where added values of lower variation or 'span' are well established. It is essential to focus on the variation control of sub-systems and individual processes. However, the Six Sigma theme of variation control has often caused the focusing of an improvement plan on a relatively narrow sector or department, as in Motorola (www.ariacad.com, December 2001):

> *Bob Galvin, former CEO of Motorola, has stated that the lack of an initial Six Sigma initiative in non-manufacturing areas was a mistake that cost Motorola $5 billion over a four-year period.*

The success of Six Sigma within General Electric was further enhanced by moving from a 'quality focus' to a 'business focus' and extending the initiative to its financial services area (e.g. GE Capital).

It is indicative, though not conclusive, that the maximum benefit of Six Sigma will be obtained by integrating it with other proven continuous improvement initiatives and extending the programme to encompass the total business. When that happens, then Six Sigma embraces the FIT Σ philosophy of integration. The shift becomes complete from a small sigma (standard deviation) to a capital sigma (Σ).

Figure 6.7 illustrates a mapping of the shift from Six Sigma to FIT Σ.

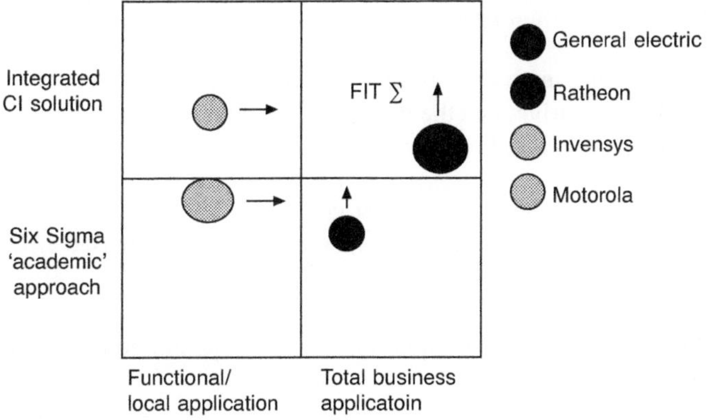

Figure 6.7 A mapping of Six Sigma.

We believe that the road map to sustainable success for the companies engaged in the 'pure play' Six Sigma programme constitutes great progress towards a company-wide integration of solutions in the FIT process.

Example 6.5 Integration of lean and supply chain with Six Sigma at Sea Gate

Seagate Technology is the world's largest manufacturer of disk drives and HDD recording media. The company's headquarters are at Scotts Valley, California, and it employs 62 000 people. Its turnover in 2000 exceeded $7 billion. The business operates in a market environment of short product lifecycle and quick ramp to high volume. The data storage market is growing by 10–20 per cent per year and the technology content doubles every 12 months. Volume products remain in production for only 6–9 months.

Seagate Springtown (which is part of Seagate Recording) started a supply chain project to improve materials management and develop a strategic vendor relationship. The fabrication plan at Springtown introduced the Lean Manufacturing philosophy that recognizes WASTE as the primary driver of cycle time and product cost. Very soon a change had taken place at Springtown and Lean Manufacturing was wholly integrated with the supply chain initiative.

The corporate office at Scotts Valley was rolling out a global Six Sigma deployment programme. The Springtown site followed the Six Sigma training programme and implemented a number of tools and techniques, including the Process Map, Sampling Plan, Cause and Effect Analysis and Control Plans, which identified a 'hidden factory'. The less visible defects of this 'hidden factory' included:

- Repeated measurements (in and out)
- Repeated chains (post- and pre-)
- Transits between manufacturing areas
- Process steps conducted in 'non-standard operating conditions'
- High rework on a process.

The Six Sigma methodology proved a key enabler for Supply Chain/ Lean Manufacturing, and the integrated programme achieved improved process capability and quality, as shown by:

- Increased throughput by 31 per cent
- Significant impact on capital expenditure due to increased efficiency of existing equipment
- Lower work-in-progress
- 80 per cent pass rate on qualifications for vacuum tools (previously 40%).

Fitness for sustainability

Managers who have steered their way through the challenges of a Six Sigma or a Lean Sigma programme over the past years are probably proud of the results and of the leanness of the operation. However, such a manager has only just begun along the path of success. In a lean programme the reduction of overhead and cutting out non-value-added activities are all excellent accomplishments, but this may be like a dieting plan to lose weight without incorporating appropriate ongoing fitness routines. In a FIT Σ programme, the sustainability of performance is instilled right through the process and not just after the implementation of the deployment plan.

The fitness for the sustainability of performance is underpinned by four key processes:

1. Performance management
2. Senior management review
   - Self-assessment and certification
   - Knowledge management.

Performance management

The point that the success of Six Sigma is highly focused on measurements, both statistical and of savings, makes performance management a logical and essential component of the programme. In the context of FIT Σ we address some relevant issues, including what we measure, when we measure and how we measure.

There is little doubt, even in the present environment of advanced information technology, that a company's performance is governed by quarterly or annual financial reports. These reports create an immediate impact on the share value of the company. The majority of performance measures are still rooted in the traditional accounting practice. Senior managers are usually driven to improve the share price, and often they have personal share options in the company. Thus the traditional accounting model of balance sheets and profit-and-loss performance statements is still being used to judge the success of organizations, although by their very nature these are backward looking historical documents. Although the information age is capable of providing real-time feedback, performance is still judged on past results.

In addition to reporting basic financial measures (e.g. sales, net profit, equity, working capital and return on investment), other traditional measures have been extended to assess customer service (market effectiveness) and resource utilization (operations efficiency). Wild (2002) argues that the three aspects of customer service – specifications, cost and timing – can be measured against set points or targets. Given many resources as input to a process, resource utilization can be measured as 'the ratio of useful output to input'. Resource utilization is cost-driven, while the objective of customer service is 'value added' to the business.

The models of financial accounting, customer service and resource utilization may also be applicable to some areas of FIT Σ – after all, they can't be ignored – but they are not key aspects of the programme.

Kaplan and Norton (1996) argue that:

> *... a valuation of intangible assets and company capabilities would be especially helpful since, for information age companies, these assets are more critical to success than traditional physical and tangible assets.*

Their model, the 'Balanced Scorecard', is illustrated in Figure 6.8.

The Balanced Scorecard retains traditional financial measures, customer services and resource utilization (Internal Business Process), and includes additional measures for learning (people) and growth (innovation). This approach complements measures of past performance with drivers for future development. The Balanced Scorecard can be applied to a stable business process following a good progress of the FIT Σ programme.

Performance management in FIT Σ should also be 'fit for the purpose', and the appropriate metrics should depend on the stages of the programme. There are three key stages of a FIT Σ initiative in the context of measuring its performance, as shown in Figure 6.9.

Focus on strategic goals

As discussed earlier, larger projects in a FIT Σ programme are selected based upon an organization's strategic goals and requirements. The viability of the project is then established based on certain quantifiable criteria, including Return on Investment (ROI). At the project evaluation stage of a FIT Σ initiative, similar criteria should prevail. Although attempts must be made to show an 'order of magnitude' of ROI data, the emphasis should be focused more on strategic goals and requirements.

The measurement process at the project implementation stage is basically the monitoring of the key factors considered during the project evaluation phase. The following six factors are suggested:

1. Determine the project's value to the business, which can be reflected in the company's overall financial performance. This factor can be applied by monitoring the savings on a monthly basis.
2. Are additional resources required? If resources are outsourced, then this cost is measured and monitored. The timescale of the project is also included in this factor.
3. What metrics are required to monitor the performance of specific large projects? Examples of this factor are Defects per Million Opportunities (DPMO) and (Rolled Throughput Yield (RTY).
4. It is necessary to monitor the impact of the project on the external market – whether there is an eroding customer service or sales revenue as a result of key company resources deployed in the project.

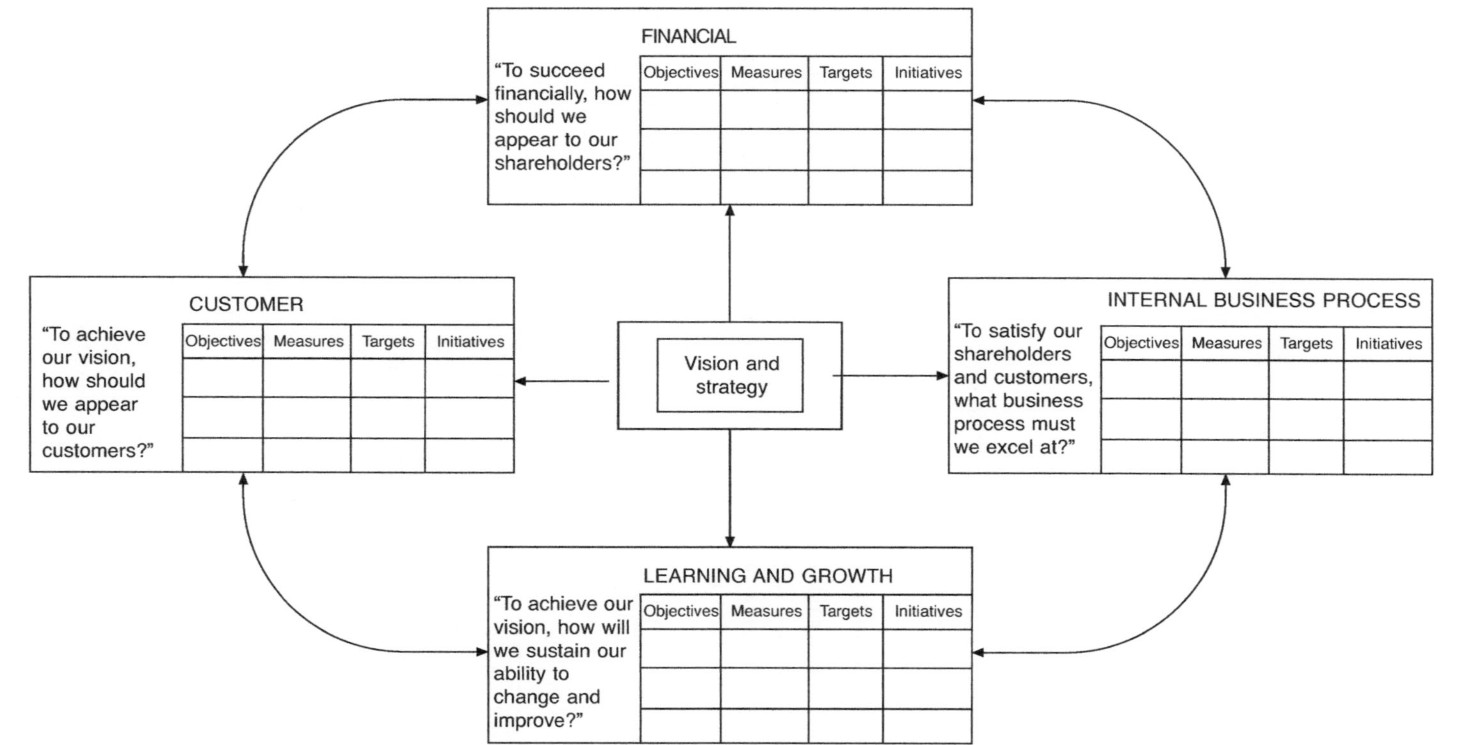

Source: Kaplan & Norton, *California Management Review*, 1996

Figure 6.8 Kaplan and Norton’s balanced scorecard.

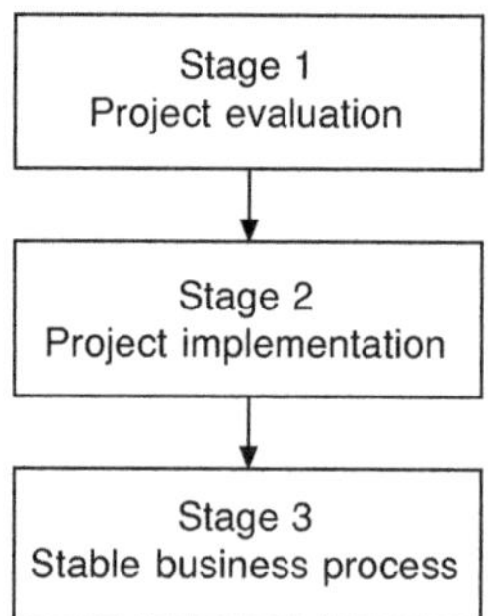

Figure 6.9 The stages of performance management.

5. Ensure that the FIT Σ initiative continues to align with the overall mission and strategy of the business.
6. Selective key performance indicators (KPIs) must be established for the next stage of the stable business process.

Transition from initiative to sustained stability

The transition period of a FIT Σ initiative from the project stage to a stable business operation is often difficult to pinpoint. The main reasons for this blurred situation are the relatively short period of project duration and the need for continuous modifications in a dynamic technological environment. Therefore five of the six factors (excluding the second factor) should be monitored for a stable business operation. However, additional emphasis should be given by focusing on the six aspects, the KPIs, and gradually all factors can then be incorporated in selective KPIs. A customized Balanced Scorecard should be appropriate for a stable business process.

Example 6.6 Monitoring performance during a Six Sigma programme at Dupont Teijin Films

Dupont Teijin Films is a global polyester films business with manufacturing sites in the USA, Europe and Asia. The company was created following the acquisition of Teijin Films of Japan by Dupont. DTF is a market leader, but experiences tough competition from new entrants. As part of the corporate Six Sigma programme, the Wilton Site of DTF in Middlesbrough (UK) started the deployment plan from 1999. The main objectives of the programme included:

- Increased capacity
- Improved material efficiencies
- Cost reduction
- Increased revenue by higher sales volume.

The site project team followed a methodology of 'successful

implementation' in three key categories – 'Doing the Right Work' (Process), 'Doing the Work Right' (Efficiency), and 'Creating the Right Environment' (Education and Culture).
Within the category of 'Doing the Work Right' the team introduced:

- Input metrics
- Output metrics
- Tracking profile.

The input metrics included the number of Black Belts trained and people trained. The Output metrics covered:

- Dollars saved
- Number of projects per annum
- Quality index
- CTQ flowdown
- COPQ
- Strategic lining.

The project team followed an internal self-assessment process every quarter, based on a 'Do Right Work Checklist' comprising 24 questions in Customer Alignment, Business Alignment, Process Baselining and Project Selection.

Senior management review

A recurring challenge for companies who have invested significant time and resources in implementing proven improvement plans such as Six Sigma is how to ensure their sustainable performance beyond the duration of a one-off corporate exercise. The annual review of the change programme during the budget planning is ineffective, because twelve months is a long time in a competitive marketplace. In order to steer the benefits of the programme and the business objectives to a sustainable future, the senior managers who are in the driving seats must have a clear view of both the front screen and the rear-view mirrors, and they must look at them as frequently as possible to decide on their direction and optimum speed.

In recent years the pace of change in technology and the marketplace dynamics have been so rapid that the traditional methodology of monitoring actual performance against pre-determined budgets set at the beginning of the year may no longer be valid. It is fundamental that businesses are managed based on current conditions and up-to-date assumptions; there is also a vital need to establish an effective communication link, both horizontally across functional divisions and vertically across the management hierarchy, to share common data and decision processes. One such solution to these continuous review requirements is Sales and Operations Planning (S&OP).

Sales and Operations Planning (S&OP) has become an established company-wide business planning process in the Oliver Wight MRPII methodology. Figure 6.10 shows the five steps in the process that will usually be present, and these are described below. The process can be adapted to specific organizational requirements.

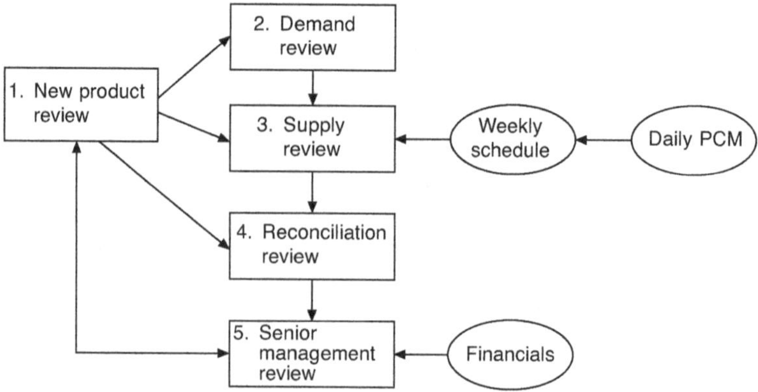

Figure 6.10 Senior management review process (S&OP)

Step 1: New product review. Many companies follow parallel projects related to the new products in R&D, Marketing and Operations. The purpose of this planning process is to review the different objectives of various departments at the beginning of the month and resolve new-product-related assumptions and issues. The issues raised will impact upon the demand plan and the supply chain at a later stage of the process.

Step 2: Demand review. Demand planning is more of a consensus art than a forecasting science. Demand may change from month to month, depending on market intelligence, customer confidence, exchange rates, promotions, product availability, and many other internal and external factors. This review at the end of the first week of the month, between Marketing, Sales, IT and Logistics, establishes agreement and accountability for the latest demand plan, identifying changes and issues arising.

Step 3: Supply review. In the current climate of increasing outsourcing and supply partnership, the capacity of supply is highly variable and there is a need to ensure the availability and optimization of supply every month. This review, usually on the second week of the month, between Logistics, Purchasing and Production, establishes the production and procurement plans and raises capacity, inventory and scheduling issues.

Step 4: Reconciliation review. Issues would have been identified in previous reviews of new products, demand and supply. The reconciliation step goes beyond the balancing of numbers to assess the business advantage and risk for each area of conflict. This review looks at issues from the business point of view rather than departmental objectives. This is also known as the Pre-S&OP review, and its aim is to minimize issues for the final S&OP stage.

Step 5: Senior management review. Senior Managers or Board Members, with an MD or CEO in Chair, will approve the plan that will provide clear visibility for a single set of members driving the total business forward. The agenda includes the review of key performance indicators, business trends of operational and financial performance, issues arising from previous reviews and corporate initiatives. This is a powerful forum to adjust business direction and priorities, and is also known as the Sales and Operations Planning (S&OP) review.

In each process step the reviews must address a planning horizon of 18–36 months in order to make a decision for both operational and strategic objectives. There may be a perceived view that S&OP is a process of aggregate/volume planning for the supply chain. However, it is also a top-level forum to provide a link between business plan and strategy. The continuous improvement and sustainability of company performance by a FIT Σ programme can only be ensured in the longer term by a well-structured S&OP or senior management review process. The results and issues related to FIT Σ should be a regular item in the S&OP agenda. Figure 6.11 illustrates how a hierarchy of KPIs can be applied and cascaded across the review processes.

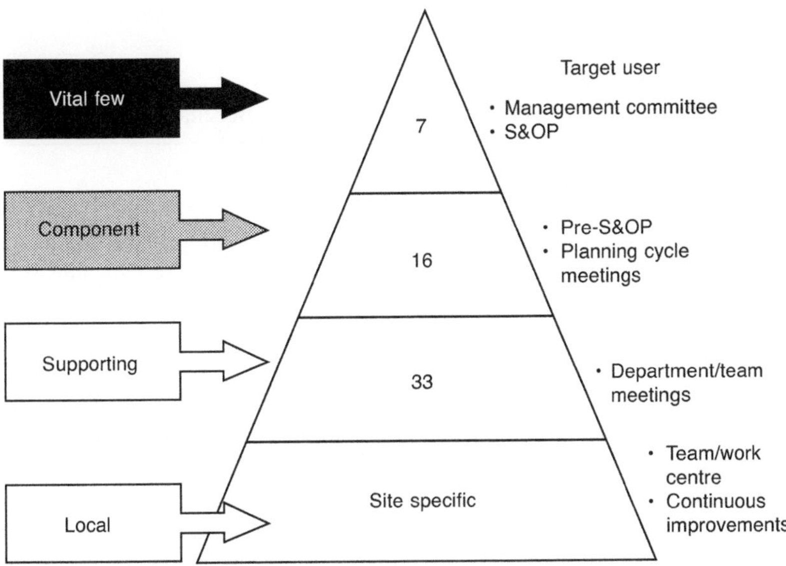

Figure 6.11 Balanced Scorecard: measures and hierarchy.

Example 6.7 S&OP ensures sustainable performance at GSK Turkey

GlaxoSmithKline Turkey (GSK Turkey, previously known as GlaxoWellcome Turkey) was awarded MRPII 'Class A' certification in 1999 by business education consultants Oliver Wight Europe.

GSK Turkey launched a programme (known as EKIP) in January 1998 to improve company-wide communications and sustain a robust business planning process using MRPII 'best practice' principles.

Since September 1998, the company has improved and sustained a customer service level at 97 per cent and inventory turnover of around 5.0. The sales turnover in 1998 increased by 20 per cent in real terms in spite of some supply shortfall from the corporate network in the first half and the adverse economic and political conditions of Turkey. GSK Turkey has been recognized as a major business in the pharmaceutical giant GSK Group, and the business plan for 1999 was aiming at a turnover of US $110 million.

As part of the MRP II Class A programme, GSK Turkey installed an S&OP process, which is underpinned by a set of business planning meetings at various levels. In spite of the GlaxoWellcome and Smith Kline Beecham merger and the corporate Lean Sigma initiative, the S&OP process has been continued by the company every month.

The vigour of the S&OP process, championed by the Managing Director, has helped the company to sustain and improve the business benefits and communication culture, especially when they were challenged by a number of initiatives in hand, including:

- Transfer of office
- Rationalization of factory and warehouse
- Corporate Lean Sigma programme
- Merger of GlaxoWellcome and Smith Kline Beecham.

Self-assessment and certification

In order to maintain a wave of interest in the quality programme and also to market the competitive advantage of quality, many companies dedicated effort to the pursuit of an approved accreditation such as ISO 9000, or an award such as the Malcolm Baldridge Award (in the USA) and derivatives of the Baldridge Award in other countries. The certification and awards have had a chequered history. After a peak in the early 1990s, the Baldridge Awards, although still prestigious in the USA, are not now as well supported in other countries as previously. The same applies to ISO 9000. Following the rush to gain the ISO stamp of approval in the 1990s, often driven by customer demand, many companies have become disillusioned by the need for external audits and by auditors mainly ensuring compliance with current procedures without providing input into improving standards of performance. A number of consultancy companies have introduced their own awards to progress an improvement programme (e.g. Class 'A' by Oliver Wight). Additionally, companies (e.g. GE, Johnson and Johnson) have started developing their own customized quality assessment process.

The need for an assessment programme

It is essential to incorporate a self-assessment process in a FIT Σ programme in order to sustain performance and improvement. There are simply two choices; either select an external certification, or develop your own checklist based on proven processes. Table 6.2 highlights the relative pros and cons of the two options.

Table 6.2 Standard accreditation or customized self-assessment

Option	Pros	Cons
Standard accreditation	Proven process Known to customers and suppliers Trained auditors and consultants available External networking	Too generic to fit business Invasion of auditors and consultants More expensive Not improvement driven
Customized self-assessment	Process ownership Customized to business needs Improvement-orientated Common company culture In-house knowledge-based Enables self-assessment	Lack of external benchmark Time to develop and pilot

There are several examples where a company achieved an external award based on a set of criteria but without improving business performance. There are also cases where, after the initial publicity, the performance level and pursuit for excellence were not maintained. If the process is not underpinned by self-assessment, then the award will gradually lose its shine. We therefore recommend that a FIT Σ programme should adopt a self-assessment process developed from proven processes, and two such processes are described below: European Foundation of Quality Management (EFQM) and 'Total Solutions'.

European Foundation of Quality Management

The EFQM award is derived from America's Malcolm Baldridge National Quality Award. There are similar accolades available in other countries, such as the Canadian Excellence Awards and the Australian Quality Award.

The EFQM Award was established in 1991. It is supported by the European Union, and the countries in the EU have their own support unit (e.g. the British Quality Foundation in the UK). As shown in Figure 6.12, the EFQM model provides a set of checklist questionnaires under nine categories each containing maximum points. They are:

1. Leadership	100 points
2. People management	90 points
3. Policy and strategy	80 points
4. Resources	90 points
5. Processes	140 points
6. People satisfaction	90 points
7. Customer satisfaction	200 points
8. Impact on society	60 points
9. Business results	150 points
TOTAL	1000 points

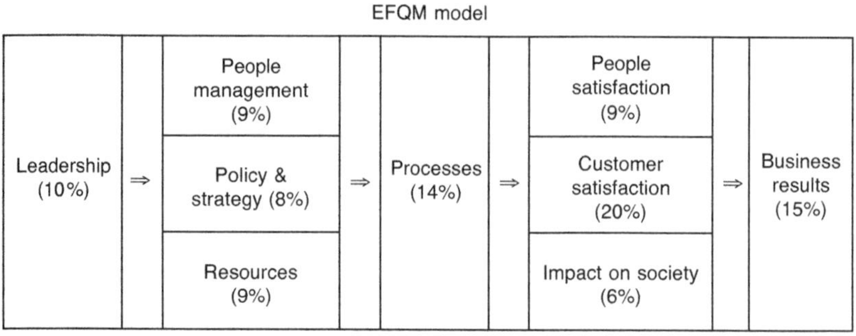

EFQM model

Leadership (10%)	⇒	People management (9%) Policy & strategy (8%) Resources (9%)	⇒	Processes (14%)	⇒	People satisfaction (9%) Customer satisfaction (20%) Impact on society (6%)	⇒	Business results (15%)

Percentages refer to the relative weighting given to criteria.

Figure 6.12 EFQM model.

The first five categories (Leadership to Process) are 'enablers', and the remaining four categories are 'performance' related.

'Total Solutions'

This holistic approach of self-analysis covering all aspects of the business has been described in detail in *Total Manufacturing Solutions* (Basu and Wright, 1998). As shown in Figure 6.13, 'Total Solutions' enables self-assessment against 20 defined areas ('foundation stones') to identify areas of improvement for achieving the full potential of the business.

The business is built from the foundation stones up, and consists of the 'six pillars' of total solutions. There are 200 questions in the checklist; ten questions for each foundation stone. These pillars are comprised as follows:

1. Marketing and innovation
2. Supply chain management
3. Environment and safety
4. Facilities
5. Procedures
6. People.

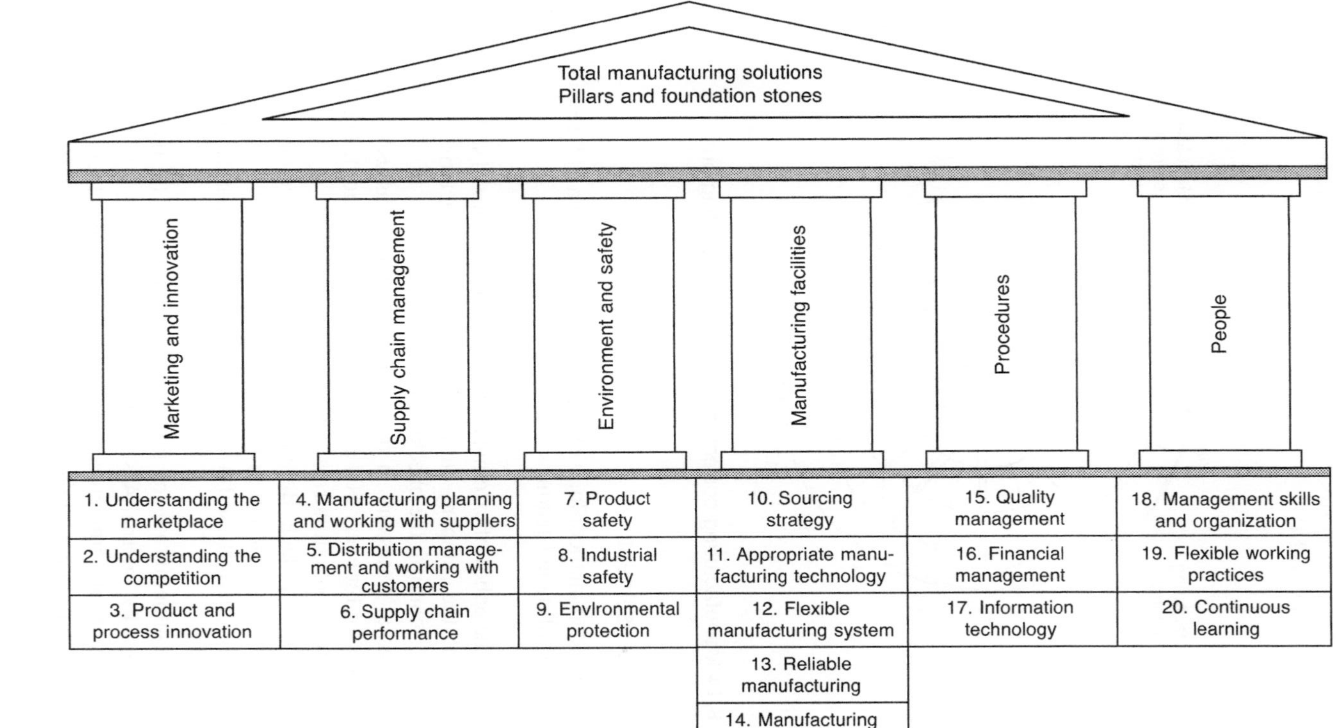

Figure 6.13 'Total Solutions': a holistic approach.

Although the checklist is aimed at manufacturing operations, it can easily be adapted to service operations. A 'spider diagram' can be constructed from the scores of each foundation stone to highlight the current performance profile and gaps (see Figure 6.14).

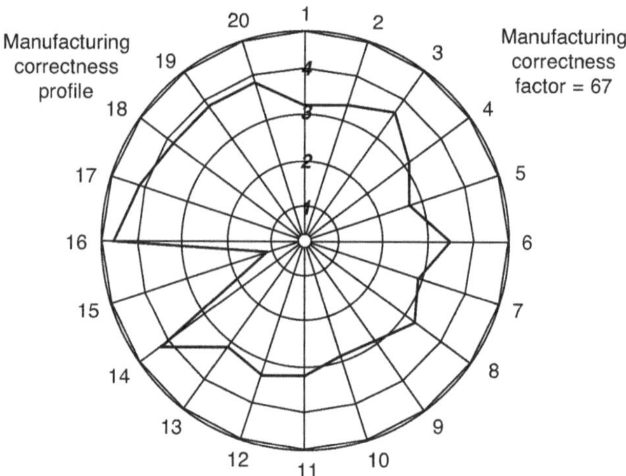

Figure 6.14 Manufacturing correctness profile (spider diagram).

FIT Σ self-assessment

Our recommended methodology of the self-assessment in a FIT Σ programme comprises the following features:

1. Establish the policy of external certification or customized self-assessment in line with the company culture and business characteristics.
2. Develop or confirm the checklist of assessment.
3. Train Internal Assessors in the common company assessment process (one Assessor for every 500 employees as a rough guide). The Assessors should also carry out normal line or functional duty.
4. Train Experts (Black Belts) and Department Managers in the self-assessment checklist and process.
5. Arrange for quarterly self-assessment to be carried out by Departmental Managers.
6. Ensure six-monthly (at the initial stage) and annual (at the later stage) assessments by Internal Assessors.
7. Analyse gaps, and implement measures to minimize them.
8. Consider corporate awards, depending on the performance attained, by the CEO.
9. Review the checklist with the change of business every two years.
10. Consider external accreditation if it adds value to the business.

The above methodology is applicable to all types of business, both manufacturing and service, and all sizes of operations, whether large, medium or small. A larger organization is likely to have its own resources to develop and maintain the process. A smaller organization may require the assistance of external consultants to develop the process.

Example 6.8 Janssen-Cilag applies 'signature of quality' for continuous self-improvement

Janssen-Cilag is the pharmaceutical arm of the Johnson and Johnson Group, and has its European Head Office based in High Wycombe, Buckinghamshire. The origins of the company lie as far back as the 1940s, with three companies initially in existence – Ortho Pharmaceutical in the UK, Cilag Chemie in Switzerland, and Janssen Pharmaceutica in Belgium. The merger was completed in 1995, and Janssen-Cilag is now among the top ten pharmaceutical companies in the world. The company markets prescription medicines in a range of therapeutic areas – gastroenterology, fungal infections, women's health, mental health and neurology.

The commitment of the company to the values and standards laid out in 'Our Credo' drives management to strive continually for excellence in a number of overlapping areas. Based upon the principles of the Baldridge Award, the Quality Management team of Janssen-Cilag developed a self-assessment process known as 'Signature of Quality' (SoQ). The process is supported by a checklist on a carefully constructed questionnaire in five interdependent areas:

1. Customer focus
2. Innovation
3. Personnel and organizational leadership
4. Exploitation of enabling technology
5. Environment and safety.

SoQ is managed as a global process from the US office, and each site is encouraged to prepare and submit a comprehensive quality report meeting the requirements. The assessment is carried out by specially trained Quality Auditors, and a site may receive a SoQ Award based upon the results of the assessment.

SoQ has been reported to be successful in Janssen-Cilag as a tool for performing a regular 'health check', and as a foundation for improvement from internal benchmarking.

Knowledge management

Almost 400 years ago, Francis Bacon stated: 'Knowledge is power'. Peter Drucker wrote in the *Atlanta Monthly*, 11 September 1995: 'Knowledge has become the key economic resource and the dominant, if not the only, source of comparative advantage'. Webster's dictionary defines knowledge as 'familiarity or understanding gained through experience or study'. In a FIT Σ initiative, an essential sustainable driver of performance management is the sharing of knowledge and best practice. Although not explicit in his fourteen-point philosophy, Deming advocated the principle 'to find the best practices and train the worker in that best way' (Deming, 1986).

The key principles of knowledge management in FIT Σ methodology are:

1. Systematically capture knowledge from proven 'good practices'
2. Select examples of the 'best practices' based upon added value to the business
3. Do not differentiate between the sources, regardless of the level of technology or economic power
4. Inculcate knowledge sharing between all units.

The essential ingredient for benefiting from knowledge management is the establishment of a 'learning organization' culture. Unless members at all levels of a company participating in sharing knowledge believe that their business can benefit from it, then the exercise has little value. If a company believes that it already knows the best way or that the 'best practice' is not appropriate to their circumstances, then sustainable improvement just will not happen. The development of a 'learning organization 'culture does not, of course, happen overnight; it takes time, and requires the appropriate infrastructure to be in place. Our experience is that time and money spend in knowledge management is also invested in the most valuable resource of competitive advantage – people. The support structure for such a knowledge-sharing process should include:

- A champion as a focal point to coordinate the process
- A regular best practice forum to learn from each other and to network
- Internal and external benchmarking to assess targets and gaps
- Continuous communication through websites, newsletters, videos and 'visual factors'.

Example 6.9 GE Capital shares best practice of 42 branches
General Electric Inc., with its global business of over US$ 120 billion per annum, has been voted by *Fortune* as 'The most respected company'. GE is also known as the 'Cathedral of Six Sigma', and the high profile of the programme under the leadership of Jack Welch has been well publicized. GE licensed Six Sigma technology in 1994 from the Six Sigma Academy, rolled out the programme worldwide, and achieved $2 billion savings in 1999.

GE Capital is the financial services arm of GE, and accounts for approximately 40 per cent of the group turnover. The success of Six Sigma in GE Capital has been a testament to the progress in service operations. One good early success story of GE Capital relates simply to the sharing of good practice.

GE Capital fielded about 300 000 calls a year from mortgage customers who had to use voice mail or call back 24 per cent of the time because the employees were busy or unavailable. A Six Sigma team found that one of their 42 branches had a near-perfect percentage of answered calls, so:

> *The team analysed its systems, process flows, equipment, physical layout, and staffing, and then cloned it to the other 41. Customers who once found us inaccessible nearly one-quarter of the time now had a 99.9 per cent chance of getting a GE person on the first try.*
> *(Welch and Byrne, 2001)*

Summary

FIT Σ is a natural extension of the third wave of the Quality Movement, offering a historically proven process to improve and sustain performance of all businesses, both manufacturing and services, whether big or small.

FIT Σ is not a statistic; it is both a management philosophy and an improvement process. The underlying philosophy is that of a total business-focused approach underpinned by continuous reviews and a knowledge-based culture to sustain a high level of performance. In order to implement the FIT Σ philosophy, a systematic approach is recommended. The process is not a set of new or unknown tools; in fact, these tools and cultures have been proven to yield excellent results in earlier waves of quality.

The differentiation of FIT Σ is the process of combining and retaining the success factors. Its strength is that the process is not a rigid programme in search of problems, but an adaptable solution for a specific company or business.

Small wonder then that FIT Σ can be seen to offer new and exciting possibilities in the field of operational excellence. There is no magic formula in a new name or brand; what counts are the underlying total business philosophy process and culture of FIT Σ.

7

FIT SIGMA™ and service organizations

The real voyage of discovery
Is not seeking new lands
But seeing with new ideas

Proust

Introduction

It is generally agreed that 75 per cent of the workforce in the United Kingdom is engaged in service industries. This high percentage is not unique to the UK; it is representative of employment statistics for developed nations throughout the world. Indeed, the US Bureau of the Census shows that over 80 per cent of the workforce in the USA is employed in service industries. Although a shift back to manufacturing industries has been identified (Basu and Wright, 1998), nonetheless it is obvious that the larger percentage of the workforce of developed nations will continue to be employed in service-type activities.

There are several reasons for this. First, continual advances in technology mean that manufacturing is considerably less labour intensive than in previous times. Automation, robotics, advanced information technology, new materials and improved work methods all have led to the decimation of manual labour. Second, for larger organizations, manufacturing has become internationalized. Many companies (for example Nike) that began with a manufacturing base are now seen as primarily marketing and service companies, with manufacturing being supplied by contractors or allied companies situated all round the world. Additionally, organizations can no longer regard themselves as being purely in manufacturing and hope to survive. The market first and foremost now demands quality of product *and* service. Market expectations of the level of quality are driven by perceptions of what technology is promising, and of what the competition is offering.

Organizations now operate in a global market where national barriers and tariff and customs duties no longer provide protection for a home market. Any manufacturer, even if it has concentrated its efforts on supplying a local market, is in reality competing on the world stage. Competition is no longer limited to other local organizations, and the fiercest competition in the home market will be from overseas-produced goods. This fact alone has meant that manufacturers can no longer make what suits their engineering strengths, but must now be aware of what the market wants and what global competition is offering. And what the competition is offering is service, in the form of delivery on time, marketing advice, training, installation, project management, or whatever else is required to provide a total service as well as a reliable product. For example, IBM, the consistent leader in the development of information technological for over 50 years, now no longer relies on selling technology, but markets solutions. This means getting alongside the customer, finding out their problems and then using the technology to provide a solution, with the emphasis being on the solution and not the technology.

Finally, it has to be recognized that customers today are better travelled, more informed and have higher expectations than did their predecessors. Much of customer expectation for continuously improved product and service has arisen from global competition, and the well publicized Total Quality Management (TQM) drive of the 1980s into service industries. Thus for over 20 years customers have been told that they are kings and queens, and to expect as a right products that work first time and a high level of service. Even government departments have mission statements espousing service, although the reality might be somewhat less than that desired by the customer!

Customers therefore expect, and take for granted, a reliable high quality product for their money. Most organizations realize that their products actually differ very little from those of the competition; any technological improvement is soon copied, and the only real difference – the 'competitive edge' – comes from service.

The divide between service and manufacturing

If no serious operation can ignore market demands for service and world-class quality, why bother to try to separate manufacturing from service? Indeed, for a manufacturing organization aspiring to world-class status we would agree, most emphatically, that the management of such organizations must concern themselves with service and quality if they are to compete on the world stage. However, a manager in a service industry such as health, retail, distribution, education, travel, real estate, consultation, brokering, law, accounting, administration of central and local government, transportation of goods or people – the list is endless – where no direct manufacturing is involved, or where the manufacturing is light and simple (such as in a restaurant), does not have to know much about manufacturing. Although all the above industries are reliant on manufacturers to varying degrees for the equipment they use, or in the case of retailers for the goods they sell, the actual physical heavy work of making the goods is not their concern. The analogy is that of a driver of a car – it is possible to be a very good driver without knowing very much about what happens under the bonnet. Some knowledge as to when to change gear, and the danger of overheating due to lack of oil or water will be of advantage, but not much more is really necessary. Likewise for management in service industries; a detailed knowledge of line balancing for a high-tech mass production assembly line of washing machines is not necessary for a retailer of white wear. Some knowledge of lead times for deliveries, operating instructions and the capacity of the washing machine is sufficient for the sales person as a basis for good advice and service to the customer.

Thus there can be a separation of management into two distinct streams; the first is the management of production operations (including service), and the second is management of operations in service industries where only some rudimentary knowledge of manufacturing is required.

The first stream, the manager that is involved directly in production and manufacturing, needs to be well versed in strategies, tactics and methodologies of production management, and also has to be very aware of what constitutes service and quality from the customer's point of view so as to be able to provide a quality product coupled with the service required to better the competition.

The second stream consists of managers who are primarily engaged in service industries. These people do not need a detailed knowledge of production systems and methodologies. They need to concentrate on providing a better service than the competition, and on using their resources as efficiently as possible to provide this high level of service.

The theme of this chapter is the elimination of non-value-adding activities and the provision of customer satisfaction in service industries using FIT SIGMA. A whole systems approach, from supplier and subcontractors through the service provider to the customer and back again, is taken.

Definition of a service organization

A service organization exists to *interact* with customers and to satisfy customers service requirements.

A service organization is when two or more people are engaged in a systematic effort to provide services to a customer – the objective being to *serve* a customer. For any service to be provided, there has to be a customer. Without a customer, and *interaction* between customer and the service organization, the objective of providing service cannot happen. The degree of intensity of interaction between the customer and people of a service organization varies, and depends on the type of service offered. For example, a specialist medical consultant will have a high degree of 'face-to-face' interaction with the customer, and so will a hairdresser. Further down the scale of face-to-face interaction comes a restaurant, where the customer judges the quality of service by the level of interaction (knowledge of wine waiter, attentiveness of waiting staff) as well as by the standard of the goods provided (wine and food). The restaurant in turn will, however, have a higher degree of personal interaction than does a fast food takeaway (see Figure 7.1). At the bottom of the scale in Figure 7.1 is automatic 'cash-point' banking, where customer interaction is purely with a machine.

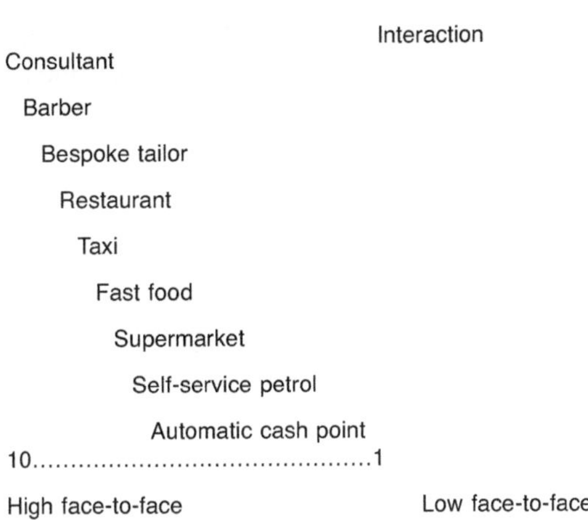

Figure 7.1 Level of face-to-face interaction.

Irrespective of the level of face-to-face interaction, without some customer interaction service cannot be provided. Note that this does not mean the customer always has to be present when the service is being provided. For example, when a car is being worked on by a mechanic the owner of the car need not be present, but nonetheless without the owner providing the car and giving instructions ('grease and oil change, ready at five?') no service can be provided.

Service operating system

A service operating system is the manner in which *inputs* are organized and used to provide service outputs.

As shown above, without a customer the objective of service cannot be delivered and therefore the customer must be regarded as input into the system providing the service.

Other resources which are inputs into the system include:

1. *Materials*. Materials used by the operating system include utilities such as energy, water and gas. Materials also include goods that are consumed by the system, goods that are transformed by the system, and goods held for sale. (Transformation refers to changing the shape or form of inputs to produce an output – for example, by placing lettuce leaves, a slice of ground beef and a slice of tomato between two halves of a bun we have

combined and transformed several goods to produce a new good, commonly known as a hamburger.)

2. *Machines/equipment.* Machines and equipment include computers, communication equipment, plant, fittings, vehicles, display racks etc., and real estate property available to the operating system.
3. *People.* This means not only the number of people employed in the operating system, but also includes the quality of the people (their knowledge and skill levels, and dependability and attitude).

All of the above represent either a capital investment or an ongoing expense to the organization.

Resources can be tangible or intangible. Tangible resources are physical, and the amount or rate of use can be measured in quantifiable terms. Intangible resources are more difficult to measure. These include:

- Time
- Information
- Attitude and skill of people.

Delivery on time is an obvious measure of performance, but measurement of effective use of time and information is less easy. Attitude and reliability of people is a value judgement, which is even harder to measure. Nonetheless, the amount of time and information available will be important issues for the manager of an operation, and likewise the friendliness and helpfulness of staff will be important issues for the customer.

With today's technology, information would seem to be readily available. The concern is *knowing* what information is required, and then being able to *interpret* and *use* information so as to achieve the organization's operational objectives.

The flow of resources through a service operating system is shown in Figure 7.2. Customer and resources are brought together to provide a service output.

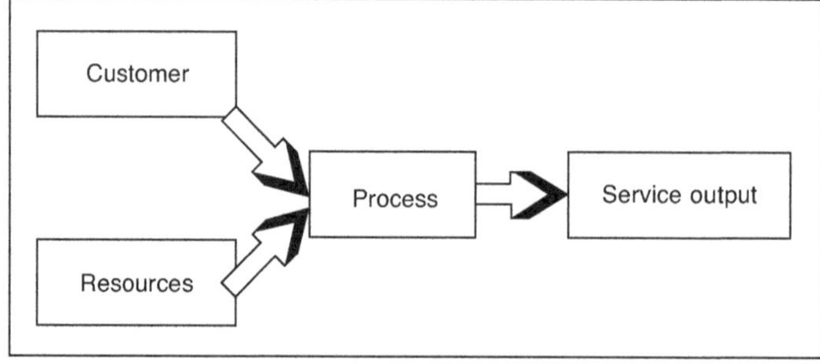

Figure 7.2 Service industry: customer driven.

Example 7.1 A bus service
A bus can travel on its advertised route, but until a passenger is uplifted the function of the bus service is not carried out. Without a passenger the function of the bus service – to carry passengers – cannot be fulfilled. An empty bus travelling on the bus route is nothing more than an unutilized, or 'stored', resource. Apart from the bus itself, resources such as fuel and time (wages of the driver) are being used.

Example 7.2 Hotel rooms/restaurants
Until a guest checks into a hotel, the service function of the hotel cannot be performed. True the room can be 'serviced' and prepared in advance, but until a guest arrives there is no service output.

Similarly, in a restaurant it is possible for the chef to make up salads, and even to prepare and cook meals in advance, before any patron is seated. This may not be the policy (strategy) of a prestigious restaurant, but nonetheless the decision (strategy) can be changed – i.e. it is not essential to have a customer before a meal is prepared. However, the function of the restaurant is not to *prepare* meals, the function is to *serve* meals, and the delivery of service cannot take place without the customer – it is not possible for the meal to be served unless there is a customer and the customer has placed an order.

In these examples – the bus travelling on its route, the prepared hotel room and the partly prepared meal in the restaurant – there are stored resources waiting for the input of a customer. Without customer input, no service output will be delivered.

Decision-making

Decisions range from long-term strategy to short-term day-to-day operational concerns. Obviously the most pressing decisions are of a day-to-day nature. Day-to-day operational decisions are limited by the objectives of the organization and by what is feasible with the resources available. Figure 7.3 shows the constraints of decision-making in day-to-day operations. The Mission /Policy of the organization sets the scope. Once the policy has been decided, then what is desirable is expressed as the objectives of the organization, and what is feasible is limited by the resources available.

The more demanding the objectives and the fewer the resources, the less choice there is. The basic objectives are two-fold – to use resources as *efficiently* as possible to achieve the highest level of *customer satisfaction* within the constraints of policy objectives and available resources.

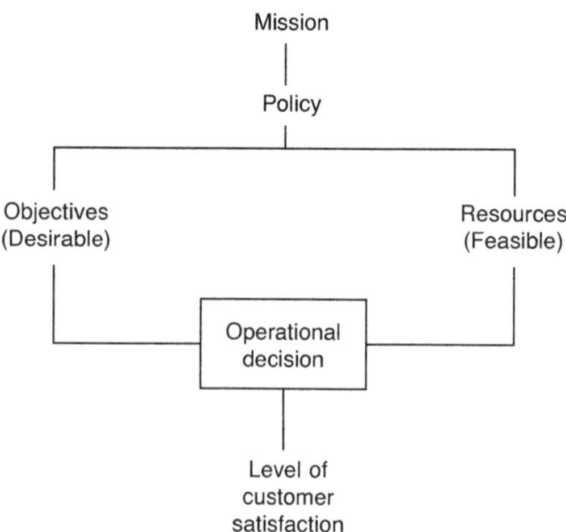

Figure 7.3 Making it happen.

Systems structure

In considering how service operations function it is useful to consider the system structure. System structure is best shown in diagrammatic form using the following symbols:

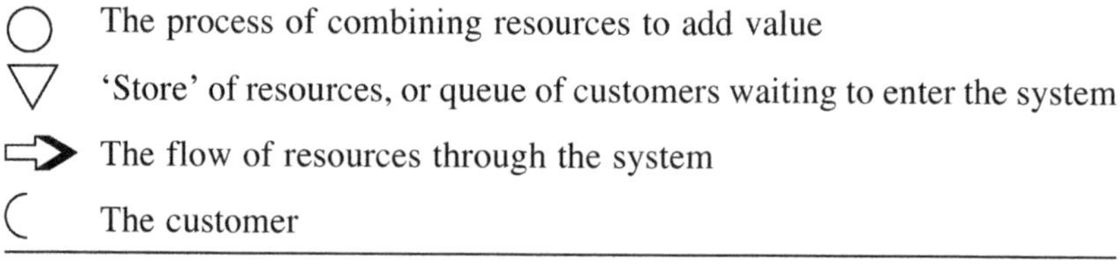

Note that the customer does not have to be external to the organization but may be an internal customer. The internal customer can be defined as the next person, or department, in the process.

Overall there are three basic service structures, but often organizations will consist of a combination of systems. The structures and the form of notation have been adapted from Wild (2002), who shows seven basic structures.

Manufacturing structures

Four of Wild's structures relate to manufacturing operations, where the customer is seen as pulling from the system. With a basic manufacturing system there will be an input stock of materials; the operation will be to transform the inputs into finished good (outputs), and the customer will be supplied from the stock of finished goods. Wild shows this as:

▽ ⇨ ○ ⇨ ▽ ⇨ (

If finished good are not stocked and the finished product is delivered direct to the customer, then the structure will be:

▽ ⇨ ○ ⇨ (

With a just-in-time manufacturing system (see Chapter 5) and Lean Sigma the aim is not to hold input stocks, and finished goods are delivered direct to the customer. The structure for this is shown as:

⇨ ○ ⇨ (

The fourth manufacturing structure shown by Wild refers to the situation where no input stock is held (such as in food processing) and the raw materials go straight into production. This structure is shown as:

⇨ ○ ⇨ ▽ ⇨ (

With manufacturing structures the important point is that even in just-in-time system manufacturing, although it is highly desirable to have a customer order before beginning manufacturing, manufacturing can still take place without customer input. For example, Toyota in Japan makes cars in batches of one to satisfy specific customer orders, but to balance the line it is still able to make basic models for stock.

Service structures

With a service system it is simply not possible to provide a service without customer input. In simple terms, without the presence of a customer a hairdresser cannot cut hair.

Structure 1

Figure 7.4 shows service being provided direct to the customer from a stock of resource. The stock of resource could be; a bus moving from stop to stop, an accident ward waiting for patients, a restaurant waiting for diners, an accountant waiting for clients, a call centre waiting for callers, a hairdresser waiting for a customer. In this structure the customer does not normally wait, the resources do. This means that the policy has to be for sufficient capacity to be available so that no customer queues form.

Structure 2

Figure 7.5 shows how most service providers would like to operate. Customers form a queue for services, and the service provider does not carry surplus

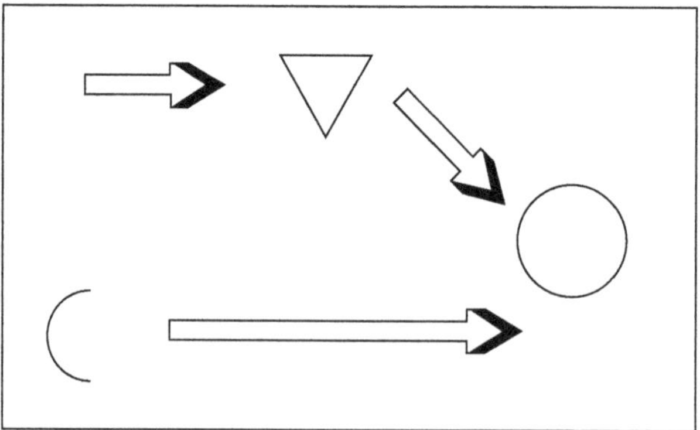

Figure 7.4 Service structures: Structure 1.

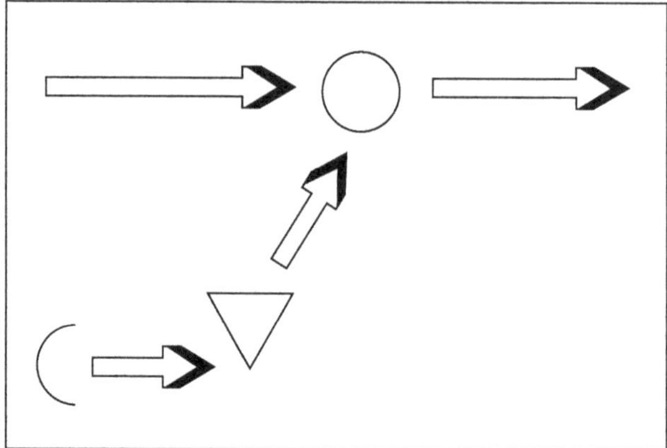

Figure 7.5 Service structures: Structure 2.

capacity – an example is a dentist or any other provider of specialist services. Customers telephone in for service, and are given an appointment. The service provider then schedules appointments – in effect, patients form a stock of customers waiting for their turn and the dentist has no idle time, unless he or she has scheduled time out. Evidence of this type of system can be seen at banks, post offices and supermarkets, where customers queue for service and the serving staff are kept busy. Refuse collection is another example; the customer puts the wheelie bin out and the bin 'waits' to be emptied. Another example is the call centre, where customers ring and are put in a queue (most government departments).

Structure 3

Figure 7.6 shows a structure where there is spare resource waiting for customers and customer queues also exist.

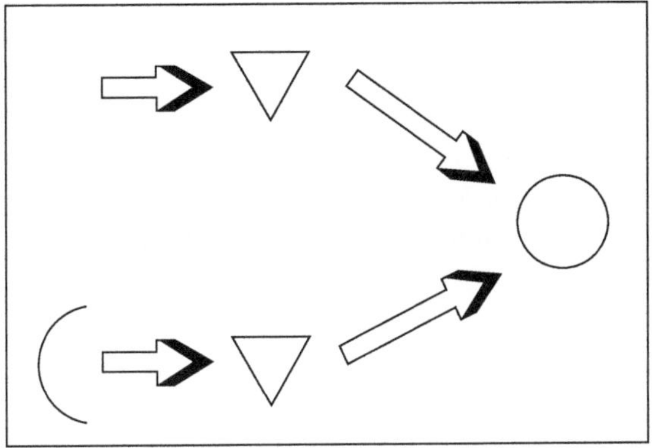

Figure 7.6 Service structures: Structure 3.

In reality, Figure 7.6 depicts how the dentist, the accountant, the lawyer, the taxi driver, the travel agent and the social service worker operate for some of the time. In effect, at times there will be spare capacity and at other times customers accumulate in queues. Additionally, although the dentist would prefer not to wait for customers but to draw customers from the accumulated stock in the appointment book, in practice he or she will be obliged to set aside some time each day for emergency treatment. If no patient arrives during the time set for emergency patients, then the dentist becomes an unused resource. If however, the norm is for customer queues and at the same time there are idle resources, this is an inefficient operation. An example is that of taxis waiting for customers at city taxi ranks, and customers waiting for taxis at the airport. The customer is frustrated waiting at the airport, and at the same time there are idle taxis at the other side of town, thus neither are resources being used efficiently and nor are customers being satisfied.

Limiting resource

In using the system structure to analyse an operation, it is important to identify the key limiting resource. For example, a hospital might be limited in the service it can provide by the number of beds available, or it might have sufficient beds but not enough staff. A school might have empty classrooms and a waiting list of students, but be limited by the number of teachers. In the case of the school with the empty classrooms and a waiting list of students, the limiting resource is the number of teachers. We would assume that in such a case the teachers are fully rostered, and thus the system structure would be Structure 2 (Figure 7.5).

Some readers will ask why a service system could not be 'just in time' – i.e. no spare resource and no customer queues. The answer is that if customers are never to be kept waiting there has to be spare resource; alternatively, if resources are going to be fully utilized with no idle time then there has to be a stock of customers to be drawn from. From time to time a perfectly balanced system might appear to exist, but this will only be a temporary phenomenon.

Combined systems

Although three basic service system structures are shown above, for most organizations a combination of structures will exist.

'Pull' systems

The structures we have looked at so far are customer 'push' systems. Another type of structure is a customer 'pull' system, which occurs where activities take place in anticipation that eventually a customer will arrive. Sometimes the projected demand is known with a fair amount of certainty in advance, and activities can therefore be safely scheduled. In this scenario the expected demand 'pulls' the system, rather than waiting for direct customer input to push the system.

Example 7.3 A motel

A small motel consists of 20 rooms. Occupancy varies, but is on average 80 per cent. Some guests book in advance, but the motel relies mainly on passing traffic. Each day previously occupied rooms are cleaned and the linen is changed in anticipation of guests arriving. The system structure can be depicted as in Figure 7.7.

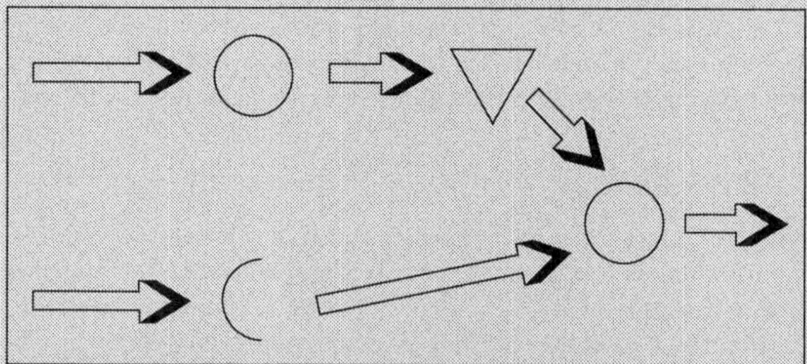

Figure 7.7 System structure of a motel.

In Figure 7.7, 'Oi' is the preparation of a room, and 'V' represents that cost has been incurred and resources transformed and held – 'stored' – in anticipation of a customer. The service operation doesn't actually occur until 'Oii', i.e. a guest arrives ('C') and a room is allotted.

Figure 7.8 is a further example of customers pulling from the system.

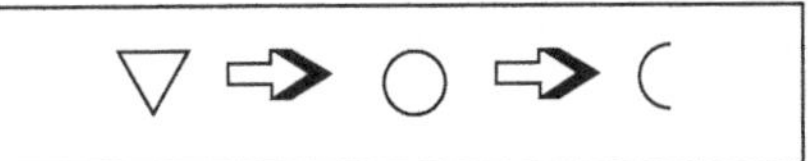

Figure 7.8 Customer pull system.

In this case the resources, such as goods for sale are stocked and the customer draws from a stock of goods. The function is the making of the sale.

Importance of structures

It is important to realize which structure(s) applies to your organization. The structure employed will determine what is feasible, and an understanding of the structures that are in force will enable consideration to be given to changing structures so as best to meet the aims and objectives of the organization. For example, it will be a policy decision that:

- Resource will be stored in advance of customer requirements (see Structure 1, Figure 7.4). Such a structure requires some surplus capacity in the system, OR
- No surplus *key* resource will be held and it is accepted that customers will queue for service (see Structure 2, Figure 7.5).

Key objectives

Managers of service operations have two key objectives. The first is to satisfy customer wants – without customers the organization will cease to exist. The second key objective is the efficient use of resources. If an organization cannot afford the level of service it is providing, it will soon go out of business. Therefore the twin objective must be the provision of customer satisfaction together (simultaneously) with the efficient use of resources.

Competition

The quality of product and the level of service provided in a competitive market must at least equate to what the competition is providing or is perceived to be providing. Customers' expectations are influenced by what they have previously experienced, by what the competition is claiming to provide in advertisements, by what the media is saying, and by the promises of technological improvements.

Basic service requirements

When introducing the concept of customer satisfaction, it has to be understood that the basic requirement for customers is that the service must first match their specification, and secondly meet time and cost constraints.

Specification – providing the customers with what they expect to receive or are prepared to accept – is the essential requirement. The service also has to be provided at a time that is acceptable to the customer, and the price must be reasonable.

What is acceptable or reasonable will always be open to question, and will depend on how important the service is to the customer and the alternatives available.

Example 7.4 A commuter bus service
Consider a commuter bus service – if a bus is not going from near where we live ('a') to near where we work ('b') then we will not catch it, if the timing does not get us to work on time we will not use it, and if the cost is too high we will seek alternatives. Thus the essential dimension of customer satisfaction is specification (the bus must be going from 'a' to 'b') – if the specification is not right, time and cost are irrelevant. Usually customers will accept (or tolerate) a service that does not perfectly meet their requirements. The amount of tolerance will be dependent on what the competition is offering or, if there is no immediate competition, on what the alternatives are. Customers might be prepared to trade some specification for cost or timing – for instance, the passenger may be prepared to walk an extra block to catch a train, rather than take the bus, if the train fare is considerably cheaper.

Provider's perspective

From the perspective of a services provider, what is provided has to be what can be afforded, and it must be at least up to the same standard as the competition. The determination of what to provide is based on economic considerations rather than altruism. Customers are needed for income, but in the long term the organization cannot afford to run at a loss. Many an organization has failed to survive although it has provided customers with excellent service.

Having offered a service that attracts customers – the specification is near to what they want, and at an acceptable time and price – we then need to look at ways of improving the service. The best protection against new competitors entering the market is to be so good that they are discouraged from entering.

Added value

Generally, some added service can be provided at very little cost. Using our bus service as an example; assuming that specification, cost and timing meet the customer's basic needs (the bus is going to the right place at the right time and the price is right), additional quality service attributes that would probably be appreciated by the passengers might include punctuality, cleanliness, a friendly, well-presented driver, and consistency of service. Achieving a punctual service is achieved by good planning and should not cost the company any extra; keeping the bus clean might add marginally to the cost (cleaning materials and wages); issuing the driver with a smart uniform will obviously be a cost; and training a driver to be courteous and well groomed might also incur costs. Although all such costs are minimal when compared to the overall operating cost of a bus company, the overall perception will be of an improved service. However, the basics of specification, timing and costs have not changed.

Once a service level has been established, then the standard must be maintained. It is important to remember that, above all, customers expect a reliable and a consistent service. A service that is sometimes excellent and sometimes indifferent will only confuse the customer.

For any organization, increased service at little or no cost will require a special culture. The workforce has to be enthusiastic and must have some authority to make limited operational decisions. Creating a quality culture resulting in staff motivated to reduce inefficiencies and to give friendly and consistent service is essential for Six Sigma, and for FIT SIGMA.

Reverting to our bus service, having achieved the basics – right route, right time, right price, clean bus, friendly well-presented driver – if the customers are now surveyed and it is found that they would also like a more frequent service, sheltered waiting areas and more comfortable buses, economics will dictate whether this is possible. Additional service at this level – a bus every ten minutes rather than every half-hour, the provision of bus shelters, and an upgrade of the fleet – adds an appreciable amount of cost, and the economics of doing so, rather than what the customer wants, will determine if such additional services are provided. However, it might be possible (if there are sufficient buses and drivers) to provide a ten-minute service at peak periods and reduce the service to once an hour at other times without adding to the overall operating cost, while at the same time providing a better service for peak-hour passengers.

We will now concentrate on determining who the customer really is, who the stakeholders are, and how to rank the relative importance of the various requirements of customers and influential stakeholders.

Who is the customer?

At the outset of this chapter it was established that in service industries the customer is an input into the process. Quite simply, without a customer no service can be performed.

Internal customers?

In Chapter 2 we discussed the philosophy of Total Quality Management (TQM). Some of the proponents of TQM consider the customer to be the next step in the operating process. For example, with TQM a writer when passing a manuscript to a word processor operator would consider the *operator* to be the customer. The TQM approach would appear to conflict with our stance, which is that in a service industry the customer is an *input* into the process rather than the *next step* of the process – thus we would show the *writer* as the customer.

The TQM concept of the internal customer was always a contrivance, initially aimed to get factory workers on an assembly line to reduce waste and pass on a good job to the next operator in the process. It was easy to say that without customers goods cannot be sold, and without sales the factory will close, but for the operator wielding the screwdriver and faced with a seemingly never-ending assembly line the customer was remote and faceless. Making the next person on the line become the customer was meant to give the customer a face. We do not criticize this approach – anything that serves to make work more meaningful, gives people more esteem and reduces costs has to be applauded. However, in reality it has to be accepted that factory workers have very little control over the quality of the product – workers does not determine the thickness or quality of the material or decide how many coats of paint will be applied, and even they tried to take a craft-worker's approach to the job, the time available to add the finishing touches to their small step in the overall process would be restricted by the speed of the line. Suffice to say that the concept of an internal customer, as the next step in the process, will not help to determine what the end user (the real customer) of the product or service really wants, and nor will it help when trying to analyse the structure of an operating *service* system.

There can, of course, be customers drawn from within an organization.

Example 7.5 A research department

A large organization with its own research department has moved to establish cost/profit centres. Previously the research department relied on other departments of the organization for research projects – the research department was not proactive in seeking work. The operating structure is shown in Figure 7.9.

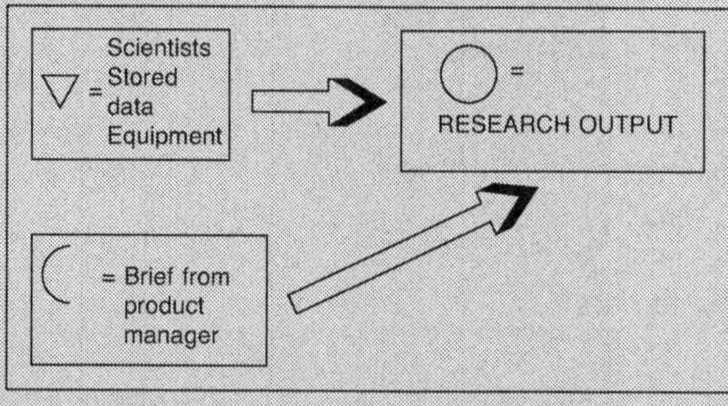

Figure 7.9 Re: active structure.

The resource of the department is the knowledge of the team, the data that has been collected and stored, test equipment, computers and so on. The customer input is the brief for the project.

Using our systems structure approach, it is apparent that without input from the customer no research will begin, and it is also likely that the product manager and staff will be consulted at various stages as the research proceeds – therefore the customer, although internal to the organization, is more than (in TQM terms) just the next step in the process. To limit the department to being the next step in the process does not encourage proactivity, but encourages the department to think of itself as the 'customer'. As a 'customer' the culture is NOT to go looking for work, but to wait for work to come to it.

Once it is realized that the customer is an input, rather than the next stage in the process, it can be appreciated that the department cannot afford to wait passively for briefs from other departments. To survive, the research department has to be proactive. It needs to promote itself within the organization, and if sufficient customers are not available from *within* there should be no reason why it should not promote itself *outside* the organization – e.g. look for work external to the organization.

Example 7.6 A pension fund department

A pension fund department of a water board is now seeking to manage funds for outside companies. A few years ago this would not have been contemplated by the department, and even if such a suggestion had been put forward – 'to go outside the organization for pension customers' – it is likely that it would not have been sanctioned by the board. This is just another example of how deregulation and privatization around the world have set the environment for government-type organizations to become commercially orientated.

Satisfying the stakeholders, or who pays the ferryman?

A stakeholder is anyone who has an interest in what an organization does. This might seem a very broad definition, and indeed it is. Knowing who stakeholders are and how their concerns might affect the operation of an organization is critical.

With some organizations, usually public sector-type operations such as education, health and social welfare, the person with the direct input into the system (the student, the patient, the welfare beneficiary) has to be satisfied. Without these people, the direct customers, the need for the service will disappear; however, in many cases the direct customer does not personally pay for the service received. Funds for the operation come from the government, or in some cases charitable trusts, and the body that provides the funds obviously has a *stake* in the efficiency of the operation. These stakeholders, the fund providers, should and increasingly do seek value for money. Value for money not only includes providing a level of service to the customer but also includes efficient use of resources. There are also other stakeholders who do not directly provide money, such as the general public in the guise of taxpayers, who are also concerned that their money is being spent wisely.

Example 7.7 A government-funded university

For a government-funded university, the customer is the student (the direct input into the teaching process). A major stakeholder is the government (main source of funds), whose chief concern is getting value for money. Other providers of funds include fee-paying students and their sponsors (parents or work organizations), and business houses who make grants or sponsor a chair. All these stakeholders will have a stake in the *quality* of the outputs. Other stakeholders who do not directly provide funds for the university but who have a very real interest in the quality and relevance of the teaching provided (for example, relevant professional bodies/societies) may have some direct say regarding what is taught for law and/or accountancy degrees. Likewise others who may not directly contribute funds, such as some parents, employers and prospective employers of graduates and, finally, the staff of the university, are all stakeholders. Each group of stakeholders is likely to have different priorities in judging the service provided; some, such as fund providers and taxpayers, will be anxious that resources are being efficiently utilized (money is not being wasted), while others will be more concerned with what is being taught, and the value of qualifications (the perceived standard or status of the university).

Determining stakeholders

For commercial businesses a stakeholder is anyone with a pecuniary interest in the organization (such as shareholders, banks, financiers, investors, suppliers of goods and services, and the people who work in the organization and their families). Other more general stakeholders include the share market, local bodies in the district of the operation, people who live and work in the operation's general neighbourhood, and the Green movement. For government and quasi-government organizations, charitable trusts and other like bodies, stakeholders are fund providers, bankers, suppliers, people who work in the organization and their families, and the community at large.

Customer/stakeholder priorities

Customer satisfaction therefore has two elements: first, we have to *know* exactly what the customers want in terms of specification, price and timing; and second, we also have to ensure that what is being offered and the manner in which we operate to satisfy the customer is not conflicting with the interests of stakeholders.

Even if there are no stakeholders, and there are only customers to be satisfied, it is important to determine different groups/segments of customers. Pareto analysis will be a useful tool, and it may be found that a vital few will account for up to 80 per cent of the business.

Stakeholders such as banks and creditors (suppliers of goods and services) are generally only interested in the financial security of the business. Other stakeholders, such as people living in the neighbourhood of the operation, have other concerns such as pollution, noise, and perhaps even heavy traffic flows. If local concerns are known in advance, then action can be taken to prevent offence. Actions that have to be taken as a result of protests or legal initiatives not only taint an organization's reputation, but are also be more costly than if the operation had been set up correctly, and stakeholders' concerns had been known and addressed in the first place.

Composite customer service rating

Christopher (1992) gives a method of rating customer service, and this is illustrated in Table 7.1.

Table 7.1 Composite customer service rating

Service index	Weighting (%) (a)	Performance (%) (b)	Weighted score (a × b)
Order fill	45	70	0.315
On time	35	80	0.28
Invoice accuracy	10	90	0.09
Returns	10	95	0.095
	——		———
	100		0.78
Composite Customer Service Rating			*78 per cent*

In the example in Table 7.1, the key criteria has been established as order fill and has been given a rating of 45/100; on-time delivery is the next most important, and other important criteria (but of lesser rating) are invoice accuracy and the number of returns (returns represent faulty goods). Column 'b' shows that 70 per cent of orders are filled, 80 per cent of orders are sent on time, the accounts department are 90 per cent accurate, and 5 per cent of goods are faulty. Christopher's composite customer service rating is calculated against internally set standards of service, and is calculated on internally gathered data, rather than on feedback from customers. With FIT SIGMA, our approach would be to ask the customers what they rate as most important and then set up a project team to establish how we could achieve 100 per cent for that particular criterion.

If we established from customers that order fill was indeed the most important, and the customers rated this at 45/100 in importance, we would now aim to score 100 per cent performance. This would be the brief for our FIT SIGMA team. Once 100 per cent was achieved, then our total performance would give a Customer Service Rating of 91.5 per cent.

However, no matter how good or how relevant we think our own internal measures are, such as Christopher's composite service rating, there is no better method than to ask the customer. Ideally, internal measures should be set against targets established by the customer. It could be that what might seem trivial to the business has become, in the customers' eyes, a major problem. For example, we might find that an important customer claims that it is impossible to get through on the telephone. Once we appreciated this we could then set the target that the phone must be answered in three rings, and whoever is passing must pick up the telephone if it is unattended. Leaving a message on voicemail is a very poor second best option. To stress the point that we are easy to communicate with, we could also set a target that all e-mails/faxes are replied to on the day received. No internal measures of such targets are needed if the culture of the organization is such that all the staff are driven by a desire to satisfy the customer.

Gap analysis

The level of service offered stems from the business policy, which in turn is to a large extent driven by what the competition is doing or is threatening to do. When deciding and specifying a level of service, management relies a great deal on the advice of the marketing function. If the marketing function does not correctly interpret the requirements of the customer, then there will be a gap between the level of satisfaction the organization believes it is providing and what the customer believes they are getting. The concept of service gaps arose from the research of Berry (1988) and his colleagues (Parasuraman *et al.*, 1985, 1991; Zeithaml *et al.*, 1990). As Lewis (1994, p. 237) says:

> *They defined service quality to be a function of the gap between consumers' expectations of a service and their perceptions of the actual service delivery by an organization; and suggested that this gap is influenced by a number of other gaps which may occur in an organization.*

The magnitude of the gap will be compounded by the number of steps in the service process and by the distance of the operational function from the customer.

Example 7.8 Gap analysis

Suppose that the marketing department's interpretation of what the customer wants is only 90 per cent correct. Straight away, this means that the actual performance can never be better than 90 per cent of what the customer really wants. If however business policy is such that it is deemed sufficient to provide resources to meet 90 per cent of customers' requirements (this 90 per cent will be set on the understanding that marketing is 100 per cent correct), then at best customers will now only get 81 per cent of what they want. Let us assume that the operation consists of a back office and a front office. Suppose the back office slightly misinterpret what management want and also set themselves an internal target of 90 per cent, and then further suppose that the front office is so resourced that to the best of their ability they can only achieve 95 per cent of the standard set; this means the final result will be that the customer is at best receiving only 70 per cent satisfaction. The calculation is as follows.

Customer requirement	100
Marketing misinterpret (get it 90% right)	90
Business policy sets target at 90% of 100 (but this actually equates to 90% of 90)	81
Under-resourced back office sets internal standard of better than 90% of target. Due to slight ambiguity and misunderstanding of management target, even when 92% of internal target is reached it is only 90% of what was set by management (90% of 81)	73
Front office, also under-resourced, is 95% on target (95% of 73)	69

Unless gap analysis is attempted, management will firmly believe that the overall result is somewhere near 90 per cent of what the customer wants. Each department, when queried, will fervently believe that it is reaching between 90 and 95 per cent of required performance levels.

If an organization is close to its customers and aware of what the competition is doing, then a gap of this magnitude should not happen. The larger the organization and the greater the delineation of responsibilities between departmental functions, and the further the operations function is removed from the customer and from consultation in business policy decisions, the greater the likelihood of gaps occurring between what is provided and what the customer really wants.

FIT SIGMA and customer satisfaction

As discussed above, generally an organization will aim consistently to achieve certain standards or levels of quality as determined by business policy. The decision as to the level of service to provide will be an economic one, driven by what the competition is doing or is likely to do. With FIT SIGMA the intention should be accurately to define what the customer wants, in terms of the basic requirements of specification, time and cost. Normally an organization will not be able completely to meet all the requirements of the customer, and some trade-offs will be possible. It is also wise to understand who the stakeholders are and what their concerns might be. If the culture of the organization is strong, the service level (as perceived by the customer) can be enhanced by enthusiastic and helpful staff at very little extra cost. Once customer needs have been accurately gauged, Six Sigma teams can be used to isolate key areas of required performance and to establish improvements and ongoing controls. The FIT SIGMA approach is not to be perfect at everything, but to concentrate on the areas that will give most benefit to the customer, and that will at the same time reduce cost by reducing or eliminating non-value-adding activities.

FIT SIGMA and resource utilization

> *Given infinite resources any system, however badly managed, might provide adequate customer service.*
>
> *Wild (2002, p. 11)*

Many an organization has failed to survive although the customers have been more than satisfied with what they have received. Thus customer satisfaction is not the only criterion by which managers will be judged. Customer satisfaction must be provided *simultaneously* with an effective and efficient operation. The level of customer satisfaction offered must not only be *affordable* to the organization; it must also be *consistent* and *sustainable.*

FIT SIGMA aims for the efficient use of resources, and the elimination of non-effective (non-value-adding) activities.

In service industries, the resources available will consist of a mix of the following:

- People
- Information technology
- Equipment and machines (display racks, checkout facilities, materials movement equipment etc.)
- Vehicles
- Space (offices, warehouses, display areas etc.)
- Materials ('intermediate' materials such as wrapping and packing materials, stationery etc.)
- Inventory (stock for sale)
- Time and information.

Obviously not all service industries will have (or need) all of these resources. There will never be an unlimited amount of resource, and often resources will be limited in both quantity and quality. An increase of resources will be dependent on funds available. When funds are not an inhibitor there can be other constraints – for example, we may need specialized packing material but it might be some weeks before suppliers can deliver.

Prioritizing resources

With FIT SIGMA, the above list of resources will be reduced or modified to show the three most *important* resources for the organization – i.e. those that are most necessary to satisfy the customers' essential requirements of specification, time and cost.

Example 7.9 A travel agent

For a travel agent, the three most important resources may be people, information technology, and space. Certainly stationery and other office supplies and equipment are needed, but these are of lesser concern than the three identified. Likewise with the branch manager's car – the branch manager might see it as an important resource, but it is of minor significance to the achievement of customer satisfaction.

Suppose that the travel agency has determined that it is valued by the customers for friendly service and useful advice on means of travel and accommodation, accurate bookings and ticketing, speedy service, and competitive prices and 'special' deals. This enables the agency to say that 'customer satisfaction is judged by specification, time and cost' – *specification* being advice and accurate ticketing, *time* equating to speedy service, and *cost* being competitive prices and special deals. To achieve customer satisfaction as defined in this manner, the agency needs a reliable integrated computer system that gives on-line information, communication with airlines, hotels and so on, and confirmation of bookings, tickets and vouchers. It also needs sufficient office space to accommodate several staff members and customers at any one time, and reliable, trained, well-presented and courteous staff.

Using the travel agency example, a matrix approach for customer satisfaction and resource utilization can be used as in Table 7.2.

Table 7.2 Balance of objectives

Customer satisfaction			Resource utilization		
Specification	Time	Cost	People	IT system	Space
3	2	1	2	3	1

3 = essential; 2 = important; 1 = less important.

For this example it has been established customers rate advice and accurate ticketing as most important (specification), and that they are prepared to wait for information and for tickets but they do not expect to wait more than five minutes before a consultant is available. Cost, although important, is a lesser consideration than accuracy and receiving speedy service. Having established this rating, the next step is to determine the vital resources needed to give the customers satisfaction. In this example a reliable integrated computerized information and ticketing system is essential. When the system is 'down', little can be achieved, information on prices, schedules, and availability of seats cannot be provided, bookings cannot be made and tickets and vouchers cannot be issued. A back-up 'manual' system consisting of the telephone, bound books of pamphlets and handwritten tickets has proved in the past to be not only unwieldy and slow but also expensive, owing to mistakes being made through information not being up to date and bookings being incorrectly recorded. 'Trained staff' is an important factor, but is of lesser importance than the system, for without the system the staff can do little. Space is an issue, but in this example has not generally proved too much of a problem. With a good system and well-trained staff, customers can be turned around quickly; when the system is slow or staff are inexperienced, then the time taken to serve a customer is extended and space can become a problem.

Example 7.10 A computer service bureau

The bureau writes specialized software to order. The customer satisfaction matrix showed that customers rated specification as important, time as not so important (they were prepared to wait to get exactly what they wanted), and that they were prepared to pay a reasonable amount. Thus:

	Specification	Time	Price
Required	3	2	2

But the perception of the service actually received by customers showed a gap between expectations and performance:

	Specification	Time	Price
Customer perception of service	2	1	1

It can be seen that customers were not satisfied. The software was not always to specification, time delays were unacceptable, and cost was too high (in comparison to what the competition was offering).

A self-analysis by the bureau found that the key resources were skilled people, own hardware, and software used for developing new programmes. Other resources, such as office space, materials, stationery etc., were of comparatively minor importance:

Resource analysis	Hardware	Software	People
Actual performance	3	1	3

The analysis revealed that the hardware was adequate and the staff were skilled, but the problem was with the in-house software. It had always been known that there were problems with the software, but it was also thought that the time delays had been caused by not having enough trained people – indeed, consideration was being given to increasing staff numbers. However, as pointed out by the accountants, an increase of staff would add to the costs. When the staff were asked for their opinion, they advised that delays and costly rewrites were due to software problems.

FIT SIGMA: balancing of objectives

The two basic objectives for an operations manager are customer satisfaction and resource utilization. The examples given above showed that by understanding the key requirements of the customer, it is possible to attempt a match with the resources available. FIT SIGMA appreciates that it will not always be possible to totally gain a balance between what the customer wants and what the organization is able to do. For the manager of an operation, a further restraint will be the objectives of the organization. If the objectives are driven primarily by the need for efficient use of resources, then customer satisfaction will be more difficult to achieve. As stated earlier in this chapter, given infinite resources any system – no matter how badly managed – might provide adequate service. The truth is that there will not be infinite resources, and existing resources will often not completely mesh with the achievement of total customer satisfaction. The manager of an operation will be expected to achieve adequate use of resources and a reasonable level of customer satisfaction. If the over-riding aim is to make the most efficient use of existing resources, it might mean that the service to be offered has to be rethought and re-promoted. Thus the service will be altered to meet the competencies of the organization, rather than extra resources being added to meet a higher-level service. Before any change to the specified service is contemplated, a Six Sigma project team approach could be used to find improved methods of operating and better ways of doing things using existing resource. Rather than saying it cannot be done, the FIT SIGMA approach is to look for ways to make the impossible possible with existing resources.

FIT SIGMA in service organizations: making a difference

The analysis of both the characteristics and management of service organizations as presented in this chapter has established that the FIT SIGMA approach can be effective in achieving service excellence. We have shown that a process is a process, regardless of the type of organization – manufacturing or service. All processes have customers and suppliers, input and output, and all processes show variation. We have attempted to dispel some mindsets or misconceptions that the service industry is different because it is intangible, inseparable from customers, perishable (cannot be stored) and variable (one off). We can see no reason why service and transactional organizations cannot look at their processes in a systematic manner.

So if the process characteristics and management tools of service organizations overlap with those of manufacturing, is it not logical to assume that the FIT SIGMA approach for both sectors should be the same? The broad answer is yes. The 'fitness for purpose' methodology of FIT SIGMA can accommodate the variability between service and manufacturing as much as the variability within the manufacturing sectors. However, there are areas where some differentiation may be applicable, in particular, in 'service level' and 'culture'.

Although it varies depending on the particular function, the process Σ governing the service level of a service organization is likely to be less than a zero-defect manufacturing organization. An error in a service function may cause severe financial penalty, but a defect in a manufactured component may cause loss of lives. A change from six sigma to five sigma is equivalent to having unsafe drinking water for 15 minutes every day. Service organizations are in an early stage of the learning curve of applying the Six Sigma tools and techniques that have been implemented in manufacturing many years back. This gap in experience should be recognized and reflected in a FIT SIGMA programme as summarized in Table 7.3 (explanations of tools are given in the glossary at the end of this book).

Table 7.3 Impact of lag in experience

	Manufacturing	Service
Programme start-up	Easier to convince top management	Difficult to convince top management
Appropriate tools	SIPOC, Flow Process, DOE, Fishbone, Pareto, DMAIC DFSS	Same
Learning deployment	Structured training programme and process Owners	Same
Project selection	Easier to quantify savings	Difficult to quantify savings
Implementation	Likely to have experienced 'derailers' and thus more difficult to implement	Easier to implement after top management approval
Sustainability	Senior management review Knowledge management Performance management and self-assessment	Same Specifically adapted for service

Summary

In this chapter we have determined that the prime objective of an organization is customer satisfaction through the achievement of a consistent and sustainable level of service. The determinant of the level of service to be provided will be driven by the competition and demands of customers and stakeholders. Providing the necessary *affordable* level of service management is vitally concerned with efficient and effective use of resources. Resources are generally limited in quantity and quality, and therefore there are potentially conflicting objectives: customer satisfaction and efficient resource utilization. FIT SIGMA recognizes this and provides an approach to determine what the customer really wants and how to match resources to essential customer needs. With FIT SIGMA it is shown that it is not essential to meet all customer needs but rather, by making the best of existing resource, to meet key needs.

8

Project management and FIT SIGMA™

As does a camel beareth labour and heat and hunger and thirst through deserts of sand, and fainteth not; so does the fortitude of a man sustain him through all perils.

Akhenaten, circa 1370 BC

Introduction

To a general manager, project management could appear to be straightforward – the very nature of any project is likely to mean that there is a definite goal, a budget and a timeframe. When the project is completed everyone knows the outcome – it is easy to judge if the project has been completed to specification, the cost can be computed, and it is very obvious if the target date has been met. If only general management was that clear-cut! With general management there is always more than one goal, and often goals are competing for resources. There is also no set timeframe – the business does not finish at the end of the financial year, the show goes on year after year. So, given the clear-cut objectives of projects, why do so many projects end up late and over budget? Note, it is reported that in Europe construction projects run over budget by 270 per cent.

The body of knowledge for project management is now extensive. Most of us will be familiar with the project management's obsession with time, and the various 'critical path' approaches to managing projects with milestones, early start, duration, early finish, late start, float and late finish calculations. There are several brands of software available that will calculate and show all of the above. However, as has been pointed out by Goldratt (1999) in *Critical Chain*, the calculation of float can be misleading. The apparent buffer of time can evaporate due to preset times and allocation of resource, for if time is saved in a preceding activity, resources may not be ready (or scheduled) to start subsequent activities early and thus the 'saving' is lost. Goldratt aside, the strength of project management is seen in a body of knowledge for managing time and planned scheduling of resource to meet intermediate and final deadlines, and the measurement of performance against time budget and specification.

Thus project management would seem to have plenty of tangible measures to manage progress of a project and judge the performance of a project manager. This being the case, what can Sigma (and more particularly FIT SIGMA) offer to the project management's body of knowledge?

This chapter identifies the common problems faced by project managers, and shows where Sigma can 'fit' so as to make the project manager's life (and consequently the client's life) easier.

The disadvantages of the project approach to management

Wilemon and Baker (1983) observed, after studying a multitude of projects, that 'there seems to be no single panacea in the field of project management; some factors work well in one environment while other factors work well in other environments'. This shows that even 20 years ago it was dawning on people, particularly practitioners, that the project approach has many flaws as a tool for getting things done. Unless this is appreciated the tendency is to blame project failures as poor management rather than a weakness in the concept of project management.

Common failures are the over-run in time and budget, and incomplete achievement of the original objectives of the project. With so many texts, and the body of knowledge, it is pardonable to wonder at the lack of transference of experience from failures.

Projects generally have four basic elements: scope, time, budget and quality. Subsidiary to these and cutting across all four are work breakdown (activities), milestones, responsibilities, cost estimation, control of costs, estimation of time, scheduling time and resource, controlling time, risk identification and management, controlling risk, balancing objectives, execution and control, finalization and close out, follow-up after hand over, team leadership and administration, and choice of information system. With the FIT SIGMA philosophy we take a whole systems approach to each of the four basic elements and the subsidiary issues listed above.

Scope

The literature is clear in that ideally the formal beginning of a project should be with Terms of Reference. The Terms of Reference should begin with a terse statement of the overall objective, such as 'To build a bridge at XYZ', 'To gain ISO 9000 accreditation', etc.

Unless the objective can be precisely stated in a few words, it suggests that clients or sponsors are not clear as to exactly what they want achieved. As Turner (1999) says, the definition of project is vital to its success. The Terms of Reference should include the background and the scope (specification or objectives), and identify key steps for completion of the project, suggested dates for each stage, the completion date and the budget. It is important that the Terms of Reference should state to whom the project manager is responsible. Finally, the Terms of Reference should be dated and signed off. An experienced project manager would not sign Terms of Reference until satisfied that all the points listed above are covered. Michael and Burton (1991) stress that Terms of Reference must always be signed off and formally approved on paper – 'a nod is not good enough'.

The Terms of Reference establish the overall scope, budget and timeframe – i.e. the three key elements of the project. The Brief follows the Terms of Reference, and provides depth. The Terms of Reference say what is required, while the Brief identifies what has to be done to make the project happen. The requirements of the Brief are reasonably accurate estimates of resources, key steps or tasks, and the skills required for each step. The Brief will also endeavour to establish cost, time and precedence for each step, and it is likely that it will also consider responsibilities and authority for the supply of resource. The Brief should not be limited to the above, but should include any issue that will affect the successful outcome of the project, such as establishing stakeholders. Stakeholders are individuals or groups who have an interest in the outcome of the project. Obeng (1994) says that a typical project will have supporters, and there will also be some who oppose it. He recommends asking, 'who is impacted by what this project is trying to achieve?' Once the stakeholders are identified, especially those who are not enthusiastic concerning the outcome, the seasoned project manager will seek to find what the concerns are and, if possible, to reassure dissenters or find ways around the concerns.

All of the above, especially the Brief, are based on estimates. By definition each project is unique, and seldom can any planned activity be taken as a certainty.

It is a recognized fact that many information technology-type projects are not completed as per the original Terms of Reference – indeed many are never completed at all! A well-published example is the UK government's project to introduce smart cards for social welfare beneficiaries. This project ran for several years and was finally abandoned in 2000 at a cost, according to the National Audit Office, of a billion pounds to the British taxpayers. When the project was abandoned, it was said that the first three-month target had not been achieved.

Project managers often give the following reasons for these types of failure:

- Clients didn't know what they really wanted, and
- Clients kept on changing their minds and adding extra features.

These reasons for over-runs in cost and time are likely to be very valid.

However, from a client's point of view the reasons for the changes are likely to be equally valid, and there will be a difficulty in understanding why the changes should make an appreciable difference to cost. Project managers are urged to view the project from a client's perspective:

- That the client appointed a project manager in the first place is evidence that the client did not have the expertise or knowledge to carry out a project.
- The client knew there was a problem and hired a project manager to solve it
- The client is relying on the project manager to advise and to suggest ways of solving the problem
- Once the project is under way, the client gains a better understanding of what is happening and will have ideas for improvements or changes.

For a short-term project such as building a house, no matter how carefully the plans have been drawn up to meet the client's wishes, once construction begins the client will see that an additional window would make sense, or that a door has to be moved and so on. After all, a plan is one-dimensional, and reality takes a different perspective once a plan begins to transform into a three-dimensional product. Such changes, if agreed on early enough, will not cause problems for the builder and should not add appreciably to the cost. For longer-term projects, such as the British Government's beneficiary payment scheme, over time the sponsor will not only change requirements, but the people that the project manager is dealing with will move on to be replaced by a new group. Indeed this particular project began under a Conservative government and carried on under a Labour government. It is easy to imagine how many changes there were in the personnel of the client, at all levels of management from ministerial down, each with ambitions and bright ideas!

FIT SIGMA solutions for achieving scope

First, we assume that the project has been properly constituted with appropriate signed-off Terms of Reference. This is so basic we apologize for mentioning it!

We have three main recommendations for achieving scope, and under the heading of scope planning we make further suggestions for achieving a common vision and for control.

First scope recommendation

Our first recommendation concerns the Brief. The FIT SIGMA approach is to be generous in estimating the resources and time needed for inclusion in the Brief, and then to make sure that the client understands that, due to the novel nature of projects – each is unique and each will have its own set of unexpected problems – estimates of time and money for resource are based on best guesses. It is important that the client understands that estimates are in reality only best 'guestimates'. Of course some clients will press for a fixed-cost project. If the project is relatively simple (as with building a house) and materials can be calculated and costed, this might be possible. However, even here allowance should be written in to enable the builder to recover major price increases of materials, and for other contingencies such as problems with foundations, water tables etc. It does not serve the client well if the builder goes bankrupt and walks off the job!

Second scope recommendation

Communicate with the client. All project managers have to remember that they don't 'own' the project; they are providing a service on behalf of the client. When the Terms of Reference were first written, the client may have emphasized finishing on time as being crucial. This does not give the project manager *carte blanche* authority to spend extra money above budget in trying to make up lost time when delays occur. Likewise if it becomes apparent that the specified completion date is under threat, the project manager has a duty to advise the client as early as possible.

We recommend weekly communication with client as standard practice. The communication should be as short as possible – two pages should be sufficient, printed on distinctive coloured paper so that the client can readily recognize (and subsequently find) the weekly report. The first page should give the actions planned for the previous week and the actual achievements for the week, while the second page should show the planned activities for the next week. If there is a variance between what should have happened and what actually happened for the previous week, a brief reason should be given as to why, together with an outline of the effect on the overall project. However, it is not sufficient to let the client know that the project has experienced a problem that may result in a time problem; a solution must be recommended, and this should include costs. The solution should never be put into effect unless the client agrees in advance to the extra cost. In this manner, if the project comes in late the client was made aware in advance, and was given the opportunity to approve the 'purchase' of extra resource to make up the lost time. If reporting is done on a weekly basis, and the client has been involved in every decision that results in extra cost or in time delays, then there will be no nasty shocks and recriminations towards the end of the project when it becomes very obvious that there are problems. As President Nixon found out, trying to cover up one small mistake can easily escalate until subterfuge becomes a strategic option – and once this strategy is adopted, it is almost impossible to turn back until the problem is so big that it becomes evident to everyone.

We strongly recommend that the provision of a weekly progress report be included in the Terms of Reference.

The benefits of the weekly report, for both parties, are:

- If the client is eager for the weekly report then it will reinforce to the project manager that the client is serious and a sense of urgency will be fostered
- It will enable the project manager to ask early for extra resource

- If problems are known and shared by both parties as they happen, or as they begin to emerge, then remedial action will be a joint decision, with no late shocks for the client
- The project manager will feel encouraged to be up-front and will feel less alone. Frequency and the accustomed regularity of communication will in itself help break down barriers and create a sense of togetherness.

Third scope recommendation

Use variation reports. If the client asks for a change that it would be possible to add, amend or whatever, and the project manager enthusiastically agrees, then often the changes are made with the project manager believing that the client has given a firm directive to go ahead. However, eventually there is a day of reckoning and the client gets the bill. The problem is that the client, when asking for a variation, did not imagine that there would be an extra cost. Take, for example, house construction. At an early stage a request to move a window a metre to get a better view might not seem a big effort (after all, the window exists and the house is still at the framework stage of construction), but this could well take the builder several hours of labour, for which he will charge. The culmination of several such minor changes, all at the request of the client (and perhaps even some suggested by the builder), might add up to several thousand pounds not budgeted for by the client. Our recommendation then is that no matter how sensible the suggestion and how minor the cost, for each variation to the Brief a cost variation advice should be issued to the client before the change is made.

Scope planning

Professor Rodney Turner (1993, 2000) explains the mechanics of scope management. He emphasizes that the purpose is to ensure that adequate work is done and that unnecessary work is not done. In FIT SIGMA parlance this means that the purpose of the project must be kept firmly in mind, and for every proposed action it has to be asked, is this really necessary for the achievement of the project? This requires a clear set of objectives for the project. The next stage is to determine the work that has to be done to achieve each objective. In project management terminology this is known as work breakdown, which means that work is broken down into areas of work that each achieve one of the project's objectives. It is important that the areas of work cover all the objectives, but not more.

The next level of planning is milestones. Milestones are the individual intermediate products or deliverables that build to the final objectives of the project. The benefits of milestone planning are that it:

- sets controllable chunks of work
- shows how each chunk of work is related and builds towards the final objective

- sets fixed targets
- fosters a common vision for all those involved, including contractors and subcontractors.

With the FIT SIGMA philosophy, the milestone plan is transparent for all to see:

- it enables the client to follow progress
- it provides a means of control for the project manager and will help in the monitoring of progress
- it enables team members to understand their responsibilities
- it shows the overall vision for everyone involved.

As Professor Turner says, a good milestone plan is understandable to everyone, is controllable both quantitatively and qualitatively, and focuses on necessary decisions. Turner recommends that, no matter how large the project, there should be no more than 25 milestones and no more than four result paths. Regardless of the size of the project, limiting the milestones to no more than 25 gives an easily manageable and comprehensible picture of the whole. Each milestone is made up of chunks of work. Obviously within these chunks of work there will be key areas to be monitored, and each chunk will in turn be broken down to have its own set of milestones.

With FIT SIGMA we do not recommend detailed planning for each milestone before the project begins. This would mean a very rigid approach, which would place unnecessary pressure on the project manager. Our approach is only to prepare fully detailed plans for activities that are about to start. Planning done twelve months out, much like the accountant's budget, will be out of date within a matter of weeks.

Time

It is assumed that the reader is conversant with network, critical path (CPM or PERT) scheduling methods. With computer packages it is possible to calculate and show early start, late start, baseline start, schedule start, actual start, duration, float, baseline float, remaining float, remaining duration, early finish, late finish, baseline finish, schedule finish and actual finish for each activity. Baseline refers to the original plan and should not normally be changed; the other times and floats should be upgraded as each activity is completed. It is not uncommon for people to say that the project has come in on time and on budget while overlooking the fact that the baseline has been long discarded, so that in fact the project is only coming in on time and against budget due to repeatedly changing the schedule at each review meeting.

We strongly recommend that the baseline, once set, is not changed – otherwise, how can we judge how good the origin plan was? Knowing where delays occurred or where estimates were optimistic will enable better planning for future projects.

Time is only one aspect of project management; it is not the be all and end all. Since the beginning of project management with the Atlas project in the space-race days of the 1950s, many managers unfortunately see project management as the management of time. This is not to say that time is not important – of course it is – but often achieving the scope will be more important, and sometimes keeping to budget might take precedence. Generally it is agreed that the three basics of project management are scope, time and budget, and a change in anyone of these will cause a change in one of the others (Figure 8.1). This suggests that one can be traded for the other – for example, more scope will mean more time and more cost, while less scope might mean less time and less cost. However, when things go wrong (i.e. a key activity has fallen behind schedule) the decision will be between reducing scope and finishing on time, or adding extra resource (cost) to try to make up lost time, or not changing the scope and finishing late (generally this will result in increased cost, as resources will be needed until the project is completed). Thus although time in itself is not the be all and end all of a project, a time delay might mean a change in scope and will almost certainly add to the cost.

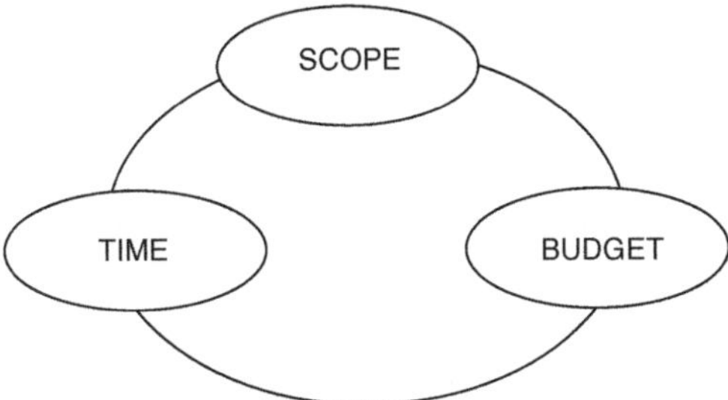

Figure 8.1 The inter-relationship of scope, time and budget.

The management of time begins in knowing the desired completion date, and working back and determining the date that each milestone must be finished by if the overall target date is going to be achieved. From this backward pass at scheduling, the amount of time available for each milestone and for each subordinate activity making up the chunks of work can be calculated. Knowing how much time is available for each activity will have a bearing on how much resource will be needed. As several tasks can be carried out in parallel, generally shown as 'paths' on a precedence diagram, it will be found that there is a float of spare time for some activities within the overall time limit of the project. Some activities will have no float, and these activities will be critical to the overall project completing on the due date. The obvious approach is to give these critical activities special consideration so that they do not fall behind schedule and delay the completion date. The problem that then arises is that if other activities are not sufficiently monitored delays can occur for these activities, and they can fall behind schedule to such an extent that they in turn put achievement of the desired completion date in jeopardy.

In the simple example shown in Figure 8.2, Activities One and Two have to be finished in sequence before Activity Six can start. Likewise, Activity Six cannot begin until Activities Four and Five are finished. Thus if both Activities One and Four begin on day one and there are no delays between activities, Activity Five will finish two days before Activity Six can start. This two days is the float for the Second Path. The First path has no float, and if Activity Six is to start on day six then it is critical that Activities One and Two are finished on time.

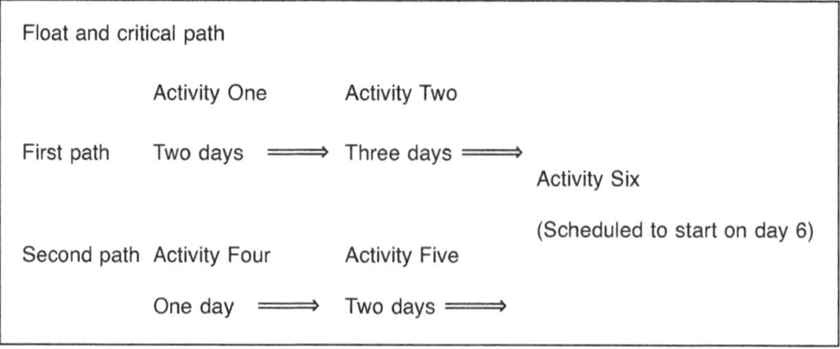

Figure 8.2 Float and critical path.

Eddie Obeng (1994) recommends allowing a buffer of time between every activity, including critical activities, when scheduling resources. He says that this buffer should be considered as a separate activity in its own right. Goldratt (1997) demonstrates how float can be lost when subsequent activities are not resourced so as to take the advantage of an early completion right through the whole project.

With these warnings and general observations out of the way, let us now consider what should be managed when considering time.

With FIT SIGMA, we recommend six steps in the control of time:

1. Set the start time, the amount of time (duration), and the finish time for each activity. Treat float as a separate activity, and where there is no float build in a buffer activity. This will become the baseline schedule, which should not be changed.
2. Do not be overly concerned with the calculations of early finish/late finish scenarios, but concentrate on the actual progress of each activity.
3. Manage and schedule resources to be available for the start date, monitor to ensure that activities finish on time, but, by having a built-in buffer, do not be unduly concerned if some of the buffer for each activity is consumed.
4. At the end of each activity, update the schedule but do not change the

baseline. If a buffer has not been consumed, move all activities forward and schedule resource accordingly. Add unused float or buffer to subsequent buffers (unless already behind the baseline).
5. If an activity falls behind schedule and the buffer or float is in danger of being used up for that activity, make the client aware of the situation, and the likely effect on whether the final baseline is going to be achieved.
6. After consultation with the client, agree on remedial action when delays occur that will endanger the overall finish date.

Budget

In the preceding section we discussed working back from the desired finish date to determine time available for activities, and allocating resources accordingly to meet the time schedule. Resources cost money, and thus in working out the time schedule we are at the same time working out the cost. As generally there will be a budget and a time line in the Terms of Reference, these two objectives are mutually dependent and a change in one will usually mean a change for the other. In the same manner as the time baseline was calculated, a budget baseline should be set at the outset.

Assuming that the budget baseline has been set, and that this baseline includes a budget for each activity, then the control of the baseline budget is similar to the monitoring and control of any expense budget. Computer printouts will provide Budget to date, Actual to date, Variance for the whole project, and likewise for each activity and each milestone. The only problem is that computer reports are provided after the event. Even if this is only one day after the event the computer record is still an historical record, and thus the project manager is always getting budget reports after the money has been spent. The aim should be to get Budget and Actual Cost reports as soon as possible, but often it will be several weeks before invoices have been received from suppliers and entered into the computer.

The FIT SIGMA approach to controlling costs

The shortcut for a project manager is to know the big costs and the fixed costs, and to make a daily allowance for all other costs. Wages will (if keeping to the baseline) be a known and a fixed cost. The wage amount for each person used on the project should be calculated in advance to give a daily or even hourly rate. For many projects wages will be a major cost. Other major costs include subcontractors, but again their charge on a daily basis should be known in advance. The hire of special equipment might be a major cost, but again there is no reason why this cannot be calculated on a daily basis in advance. All other costs should be allowed for as one fixed figure; we call this ongoing cost. Ongoing cost can be allowed for at a daily rate based on actual costs incurred in previous projects. The ongoing cost figure should be a constant, until actual costs are reported by the accountant (usually six weeks after the event). Once the actual costs are known, it might be necessary to increase the ongoing cost daily figure. Knowing the cost of wages for each person on a daily basis, ascertaining in advance the daily cost of hired equipment and subcontractors, and having a daily allowance for all other costs, the project manager can have a good 'feel' of costs on an ongoing daily basis. It might be necessary to have an assistant to keep this record, but if the project manager concentrates on only the major costs, then a rough calculation can be made in 30 minutes each day.

If it is obvious that costs are ahead of the baseline budget, then the client should be immediately informed. If the costs are not recoverable from the client, then the project manager's senior management will need to be informed! There is little future in hoping that costs can be recovered at a later stage. Even if savings can be made, it is best to be upfront when the problem occurs, and to explain the remedial actions being taken.

Quality

Quality in projects as perceived by the client is generally based on intangibles. It is taken for granted by the client that the scope, budget and time will be met – after all, that is what they paid for, and what the project manager contracted to provide.

Thus quality from a client's perception refers to the basics of scope, cost and time *plus* the intangibles of the working relationship with the team, the ability of the project team to accommodate changes to the scope, communications, ease of transfer/implementation from project to ongoing operation, training, and follow-up service after handover. For tangible projects (such as a construction project), quality will also include a judgement on the standard of finish, the cleanliness of the site, any extras provided etc.

From the project manager's point of view, quality includes all of the above plus the costs of non-conformance resulting in delays, overtime, rework, wasted materials, idle time, putting right etc. Quality and the culture of FIT SIGMA quality has been fully discussed in the preceding chapters, especially Chapters 1–3, and is equally applicable to project management. One of the key issues for project managers is the building of team spirit and the fostering of a quality culture, with all sharing the same 'can do' philosophy.

Our final comment relates to follow-up after handover. If through poor management of the client or lack of training of operators the project does not achieve full ongoing benefits, the shortcomings will be blamed on the project manager. It is therefore in the project manager's interest for the completed project to work the way it was intended. A good project manager follows up, and provides 'after sales service'. This is nothing less than sound business practice, and can lead to further business from the client or referred business.

Summary

This chapter has discussed the key issues of project management, centring on the key issues of scope, time, budget and quality. A holistic approach is taken, as none of these four issues can be considered in isolation – a change or shortcoming in any one will have an effect on the other three. Nonetheless, as the standard approach in project literature is to discuss these issues under separate headings, we have followed the same pattern.

The chapter has taken the standard and accepted approach to project management and added a FIT SIGMA wisdom so as to make the project manager's life, and consequently the customer's life, just that much easier.

9

Implementation, or making it happen

All is change, nothing is permanent.

Heracleitus (513 BC)

Introduction

When we began to write this chapter, Ron had just returned from a Six Sigma conference at the Café Royal, London. A women delegate at the conference commented: 'A Six Sigma programme is like having a baby – very easy to conceive but difficult to deliver'. The implementation of FIT SIGMA™ and for that matter the implementation of any change programme is like 'having a baby'; the delivery of change is difficult. According to Carnall (1999):

> *. . . the route to such changes lies in the behaviour: put some people in new settings within which they have to behave differently and, if properly trained, supported and rewarded their behaviour will change. If successful this will lead to mindset change and ultimately will impact on the culture of the organization.*

Facts transfer

Then implementation of FIT SIGMA is a major change programme designed to transform an organization. This transformation can only come about if the cultural change of mindset is combined with facts transfer. FIT SIGMA is not an *ad hoc* localized improvement project; it is a holistic programme across the whole organization. Therefore the essential characteristics of a FIT SIGMA implementation programme are that:

1. It is top-led, with totally committed management, and bottom-driven (see Figure 9.1)
2. The project management discipline of scope, time and budget is employed (see Figure 9.2)

3. There is rigorous specialist training
4. There is company-wide open communication, spanning all functions
5. Savings and success are measured.

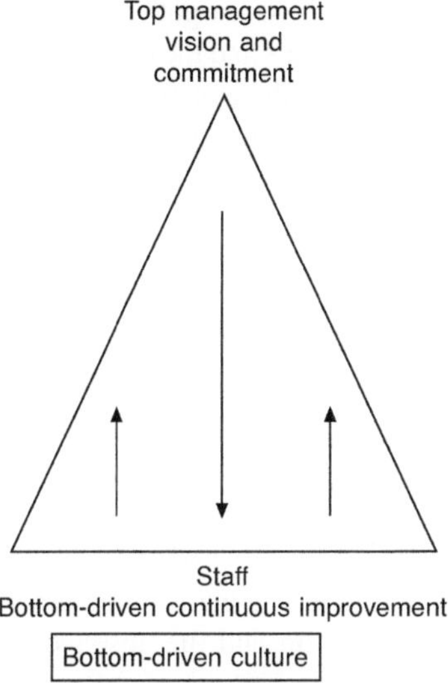

Figure 9.1 FIT SIGMA™ is a bottom-driven culture.

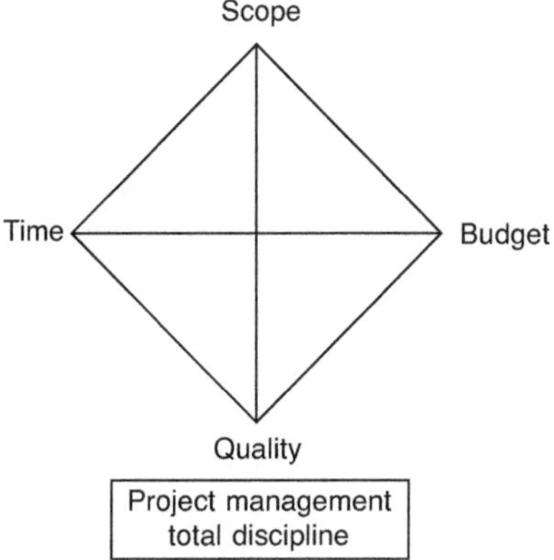

Figure 9.2 FIT SIGMA™ project management – total discipline.

In the preceding chapters we have already touched upon many factors related to implementation programmes – for example, Chapter 3 shows the four key steps of Six Sigma deployment, Chapter 6 includes FIT SIGMA methodology for improvement and integration, and Chapter 8 emphasizes the importance of project management principles. There are also many publications and articles relating to strategic change management and project management (for example Carnall, 1999 and Turner *et al.*, 1996) that it would be helpful to read before embarking on a FIT SIGMA programme. Some other literature, however, implies that if an organization follows a recommended systematic structured approach, change management is straightforward. Our experience shows that a rigid structured approach is far from a guarantee to success. We recommend the following of a proven path with some degree of flexibility, taking into account the requirements and the existing culture of the organization.

Below we outline proven pathways for implementing FIT SIGMA for organizations that are in different stages of sigma awareness and development. We have categorized three stages of development:

1. New starters of FIT SIGMA
2. Organizations that have started Six Sigma but stalled
3. Organizations that have completed Six Sigma, but where to now?

Implementation for new starters

At this stage the management understands the need for change and the need for an improvement programme. The main concern will be the change required to the culture of the organization and the absence of a proven structure for transformation of a culture. The management knows what it wants, but how does it convince the staff that they need to or want to change? You can take a horse to water, but how do you make it drink?

Here we provide a total proven pathway for implementing a FIT SIGMA programme, from the start of the initiative to the embedding of the change to a sustainable, 'bottom-driven', organization-wide culture.

Note that the entry point and the emphasis on each step of the programme could vary, depending on the 'state of health' of the organization.

The framework of a total FIT SIGMA programme is shown in Figure 9.3 and is described below.

Step 1: Management awareness

A middle manager has been tasked by the CEO with leading a Six Sigma programme in a large organization with no previous experience of Six Sigma. The CEO has just read an article concerning Jack Welch's successes with Six Sigma in General Electric, and he is full of enthusiasm and has high expectations. The middle manager has grim forebodings of failure. He realizes that the CEO is a powerful member of the board, but after all he is only one member. In another organization, the quality manager for a medium-sized company has attended a Six Sigma conference and has mixed feelings of optimism and doubt.

Our experience is that both these managers are right to be concerned. It is essential to convince the CEO and at least one-third of the Board of the scope and benefits of FIT SIGMA before launching the programme. The success

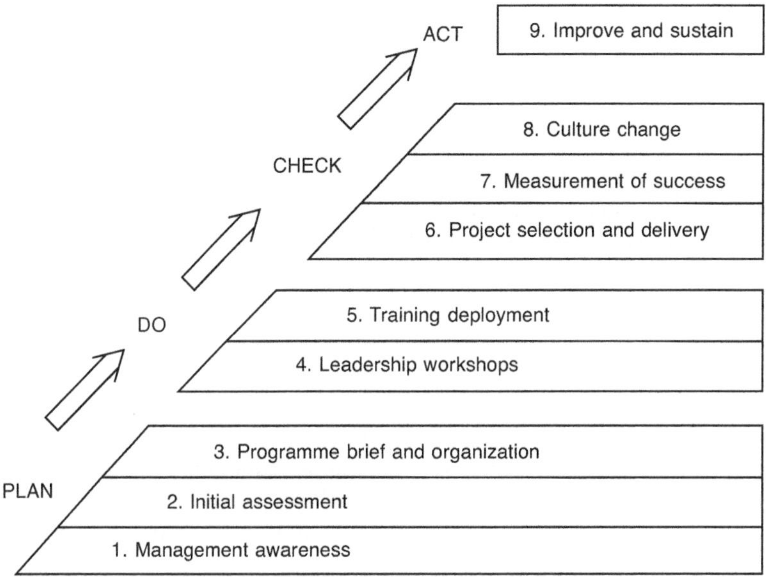

Figure 9.3 Framework of FIT SIGMA™ implementation.

rate of a 'back door' approach without the endorsement of the key players cannot be guaranteed. If a programme is not company-wide and wholly supported by senior management, it is not FIT SIGMA – it may be a departmental improvement project, but it is not FIT SIGMA. In cricketing terms, a CEO can open the batting but a successful opening stand needs a partner at the other end.

Our experience suggests that management awareness has been a key factor in successful application of Six Sigma in large organizations. Various methods have been followed, including:

- A consultant's presentation to an off-site board meeting (e.g. General Electric)
- The participation of senior managers in another organization's leadership workshop (e.g. GSK and Ratheon)
- Study visits of senior managers to an 'experienced' organization (e.g. Noranda's visit to General Electric, DuPont and Alcoa).

Small and medium-sized companies can learn from the experience of larger organizations, and indeed there can be mutual benefits for the larger organization through an exchange of visits. Service industry organizations may benefit by exchanging visits with successful Six Sigma companies in the finance sector, such as American Express, Lloyds TSB and Egg Plc.

During the development of the management awareness phase, it is useful to produce a board report or 'white' paper summarizing the findings and benefits. This paper has to be well written and concise, and should not be rushed. Allow between four and twelve weeks for fact finding, including visits, and the writing of the 'white' paper.

Step 2: Initial assessment

Once the agreement in principle from the board is received, we recommend an initial 'health check' of the organization. There are many good reasons for carrying out an initial assessment before formalizing a FIT SIGMA programme, including the following:

- Having a destination in mind, and knowing which road to take, is not helpful until you find out where you are
- Once you know the organization's needs through analysis and measurement of the initial size and shape of the business and its problems/concerns or threats, techniques of FIT SIGMA can be tailored to meet the needs
- The initial assessment acts as a springboard through bringing together a cross-functional team, and reinforces the 'buy in' at the middle management level
- It is likely that most organizations will have pockets of excellence, along with many areas where improvement is obviously needed, and the initial assessment process highlights these at an early stage
- The health check must take into account the overall vision/mission and strategy of the organization, so as to link FIT SIGMA to the key strategy of the Board; thus the health check will serve to reinforce or redefine the key strategy of the organization.

There are two essential requirements leading to the success of the assessment (health check) process:

1. The criteria of assessment (check list) must be holistic, covering all aspects of the business, and specifically address the key objectives of the organization
2. The assessing team must be competent and 'trained' in the assessment process (whether they are internal or external is not a critical issue).

It is sensible that the assessment team be trained and conversant with basic fact-finding methods, such as are used by industrial engineers. Some knowledge of the European Foundation For Quality Management (EFQM; see Figure 6.12) or the Baldridge (performance excellence) method of appraisal would be most useful.

Once the health-check assessment is completed, a short report covering strengths and areas for improvement is required. We stress that the report should be short (not the 75-page report required for EFQM). In writing the report the company might require the assistance of a Six Sigma consultant. The typical time needed for the health check is two to six weeks.

Step 3: Programme brief and organization

This is the organization phase of the programme, requiring a clear project brief, appointment of a project team and the development of a project plan – '... major, panic-driven changes can destroy a company, poorly planned change is worse than no change' (Basu and Wright, 1998).

The programme must clearly state the purpose, scope objectives, benefits, costs and risks associated with the programme. A FIT SIGMA programme is a combination of total quality management, lean management, Six Sigma and culture change management. It is a big undertaking, and requires the disciplined approach of project management – according to the Central Computer and Telecommunication Agency (CCTA, 1999), 'Programme management is the coordinated management of a portfolio of projects that change organizations to achieve benefits that are of strategic importance'.

One risk at this stage is that management might query the budget for the programme and there might be some reluctance to proceed. If this is the case, then it is obvious that management has not fully understood the need for change. This is why we have stressed the importance of the first step, management awareness. Reinforcement could however well be needed during Step 3, underpinned with informed assumptions and data including cost/benefit/risk analysis. Unless management is fully committed, there is little point in proceeding.

There is no rigid model for the structure of the FIT SIGMA team. Basic elements of a project structure for a major change programme can be found in Basu and Wright (1998) or Turner *et al.* (1996). Our suggested FIT SIGMA model is shown at Figure 9.4.

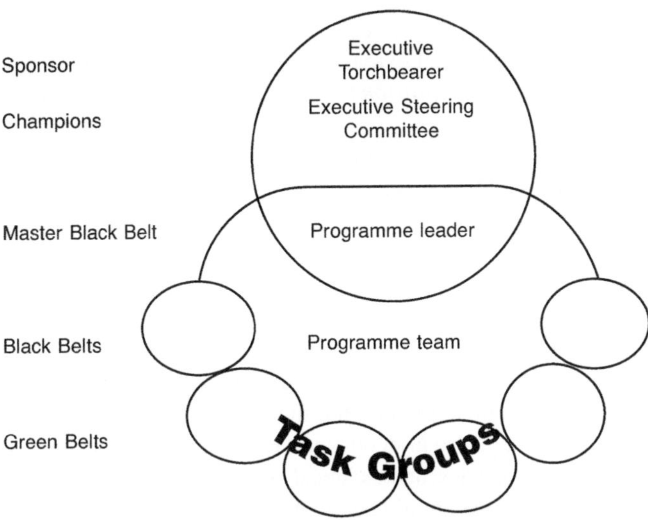

Figure 9.4 FIT SIGMA^{TM} programme organization.

Executive torchbearer

Figure 9.4 shows an executive torchbearer. The executive torchbearer will ideally be the Chief Executive (CEO), and the official sponsor for FIT SIGMA. The higher up the organization the torchbearer is, the greater the success of the programme. The role of the torchbearer is to be the top management focal point for the entire programme and to chair the meetings of the Executive Steering Committee. Being a torchbearer may not be a time-consuming function, but it is certainly a very important in order to give the programme high focus, expedite resources and eliminate bottlenecks.

Executive Steering Committee

To ensure a high level of commitment to and ownership of the project, the Steering Committee should be drawn form members of the Board plus senior management. Their role is to provide support and resources, define the scope of the programme consistent with corporate goals, set priorities, and consider and approve the programme team recommendations. In Six Sigma terminology, they are the champions of processes and functional disciplines.

Programme leader

The programme leader should be a person of high stature in the company – a senior manager with broad knowledge of all aspects of the business, and good communication skills. He or she is the focal point of the project and also the main communication link between the Executive Steering Committee and the programme team. Often the programme leader will report directly to the torchbearer.

The programme leader's role can be likened to that of a consultant. The role of the leader is to a great extent similar to Hammer and Champy's 'csar' in *Re-engineering the Corporation* (1993). The programme leader's role is to:

- Provide necessary awareness and training for the project team, especially regarding multifunctional issues
- Facilitate work of various project groups and help them develop and design changes
- Interface across functional departments.

In addition to the careful selection of the programme leader, two other factors are important in forming the team. First, the membership size should be kept within manageable limits. Second, the members should bring with them not only analytical skills but also in-depth knowledge of the total business, covering, marketing, finance, logistics, and technical and human resources. The minimum number of team members should be three, and the maximum seven; any more than that can lead to difficulties in arranging meetings, communicating, and keeping to deadlines. The dynamics within a group of more than seven people allow a pecking order and subgroups to develop. The team should function as an action group, rather than as a committee that deliberates and makes decisions. Their role is to:

- Provide objective input into the areas of their expertise during the health-check stage
- To lead activities when changes are made.

For the programme leader, the stages of the project include:

- Education of all the people in the company
- Gathering the data
- Analysis of the data
- Recommending changes
- Regular reporting to the Executive Steering Committee and the torchbearer.

Obviously the programme leaders cannot do all the work themselves. A programme leader has to be the type of person who knows how to make things happen, and one who can motivate other people to help make things happen.

Programme team

The members of the programme team represent all functions across the organization, and they are the key agents for making changes. The members are carefully selected from both line management and functional background. They will undergo extensive training to achieve Black Belt standards. Our experience suggests that a good mix of practical managers and enquiring 'high flyers' will make a successful project team. They are very often the process owners of the programme. Most of the members of the programme team are part-time members. As a rule of thumb, no less than 1 per cent of the total workforce should form the programme team. In smaller organizations the percentage will of necessity be higher, so that each function or key process is represented.

Task groups

Task groups are spin-off teams formed on an *ad hoc* basis to prevent the programme team getting bogged down in detail. A task group is typically created to address a specific issue, which might be relatively major (such as Balanced Scorecard), or relatively minor (such as investigation of losses in a particular process). By nature, task group members are employed directly on the programme on a temporary basis. However, by providing basic information for the programme, they gain experience and Green Belt training. Their individual improved understanding and 'ownership' of the solution provide a good foundation for sustaining future changes and ongoing improvements.

Timeframe

A preliminary time plan with dates for the milestones is usually included in the programme brief.

We have now covered the first three plan steps shown in Figure 9.3.

The 'Do' steps

After the Plan phase comes the Do phase (see Figure 9.3).

Once the programme and project plan has been agreed by the Executive Steering Committee, there should be a formal launch of the programme. It is critical that all stakeholders, including managers, employees, unions, key suppliers and important customers, are clearly identified. A high-profile launch targeted at stakeholders such as these is desirable.

Organization for small and medium enterprises

The structure of the programme will vary according to the nature and size of the organization; for small and medium-sized enterprises (SMEs) a typical structure is as shown in Figure 9.5. For small enterprises the programme leader might be part time; in all other cases the programme leader will be full time.

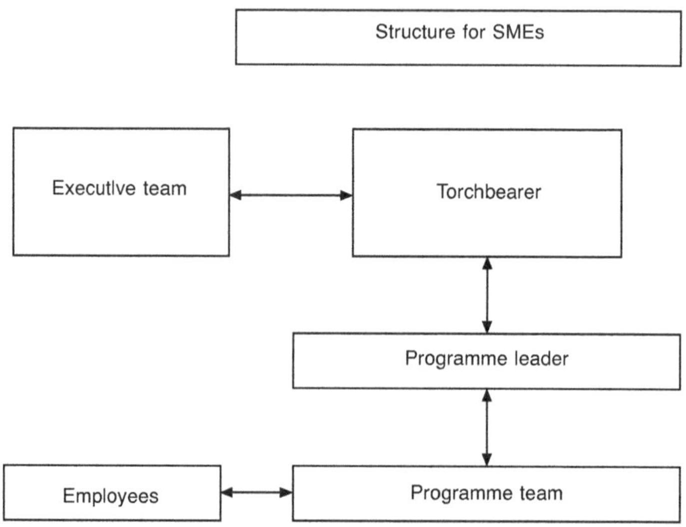

Figure 9.5 FIT SIGMA™ structure for SMEs.

Step 4: Leadership workshop

All board members and senior managers of the company need to learn about the FIT SIGMA programme before they can be expected to give it their full support and input. Leadership training is a critical success factor. Leadership workshops can begin simultaneously with Step 1, but should be completed before Step 5 (see Figure 9.3).

The leadership workshops last between two and five days and cover the following issues:

- What are Six Sigma and FIT SIGMA?
- Why do we need FIT SIGMA?
- What will it cost, and what resources will be required?
- What will it save, and what other benefits will accrue?
- Will it interrupt the normal business?
- What is the role of the programme leader and the Executive Committee?

Step 5: Training deployment

Training/learning deployment has been covered in Chapter 6 (Figure 6.2 refers). The training programme, especially for the team members, is rigorous. It might be queried whether it is really necessary to train to achieve Black Belt certification – indeed, formal certification might not be essential. However, there is no doubt that without in-depth training of key members of the programme little value will be added in the short term, and even less in the long term. The training/learning deployment creates a team of experts. It has already expected that programme members will be experts in their own departments and processes as they are currently being run, and that they will have the capability of appreciating how the business as a whole will be run tomorrow. FIT SIGMA will give them the tools for the business, as a whole, to achieve world-class performance.

It is emphasized that, apart from the rigorous training in techniques and tools, the training will also change how the members will look at things. Training is an enabler, not only to understand the strategy and purpose of change but also, as evidenced by the experience of American Express, to help members to:

- Identify project replication opportunities
- Understand leveraging the results of the programme
- Identify and eliminate areas of rework
- Identify drivers for customer satisfaction
- Leverage FIT SIGMA principles into new products and services.

Smaller organizations are very often concerned about the cost of training, especially the money paid out to consultants and for courses. In a FIT SIGMA programme, training costs can be minimized by careful selection of specialist consultants and the development of own training programmes. Porvair Limited is an example of a smaller enterprise that achieved good results with a limited training budget.

Example 9.1 Porvair Limited

Porvair is a manufacturing company based in Wrexham, England. It currently employs 80 people, and has annual sales of £6 million. There are 2000 part numbers in the product range, which is made from plastic, bronze or stainless steel, and the range is applied to porous media and filtration equipment. The customers are involved in high-temperature catalyst recovery, medical applications, nylon spinning and water filtration.

Prior to introducing Six Sigma, the company already had a respectable reputation – as demonstrated by the fact that the Welsh Development Agency designated the company as a benchmark site. However, the *ad hoc* improvement programme relied heavily on one person, the technical director. The other are of concern was that Porvair was experiencing poor delivery performance – less than 50 per cent of deliveries were on time. Following a Six Sigma programme that commenced in May 2000, by March 2002 the company had achieved remarkable results (see Table 9.1).

Table 9.1 Porvair performance increase

	May 2000	March 2002
Delivery on time	< 50%	90–95%
Head count	135	80
Customer complaints (per month)	12	7 (but now of a minor nature)
Waste	14%	10%

Plus benefits from three Six Sigma projects, £206 000.

The company deployed a specialist Six Sigma consultant from Belfast, and the costs of the training course were:

One Champion	£14 000
One Black Belt	£17 000
Four Green Belts	£14 000
Other costs	£12 000
Total	£57 000

The Black Belt (once trained) carried out further training in-house for additional Green Belts, and awareness training for all employees. The training costs at Porvair equate to £1000 per employee per annum.

Step 6: Project selection and delivery

The project selection process usually begins during the training deployment step. Project selection, and subsequent delivery, is the visible aspect of the programme. A popular practice is to begin by having easy and well-publicized successes (known as harvesting hanging fruit). This was covered in Chapter 6, where it was recommended that quick 'wins' should be aimed for ('just do it') projects.

In a similar fashion, Ericsson AB applied a simplified 'Business Impact' model for larger projects. Ericsson categorizes projects under three headings:

1. Cost takeout
2. Productivity
3. Cost avoidance.

A variable weighting is allocated to each category, as shown in Table 9.2.

Table 9.2 Project benefit categories and business impact

Level	Cost takeout	Productivity and growth	Cost avoidance
Definition	'Hard' savings Recurring expense prior to Six Sigma	'Soft' savings Increase of process capacity so you can 'do more with less', 'do the same with less', 'do more with the same'	Avoidance of anticipated cost Is not in today's cost structure
Example	Less people to perform activity Less $ required for same item	Less time required for an activity Improved machine efficiency	Avoid purchase of additional equipment Avoid hiring contractors
Impact	Whole unit	Partial unit	Not in today's cost
Weighting	100%	50%	20%

Business impact = Cost takeout + 0.5 Productivity = 0.2 Cost avoidance – Implementation cost.

When sufficient data are not available to provide an accurate estimate of Business Impact, an approach of 'Derived Importance' based upon scores for various categories is a practical alternative, as shown in Table 9.3.

For smaller, 'just do it' projects, it is a good practice to establish an 'ideas factory' to encourage task groups and all employees to contribute to savings and improvement. Very often small projects from the 'ideas factory' require negligible funding.

Project review and feedback

One important point of the project selection and delivery step is to monitor the progress of each project and control the effects of the changes so that expected benefits are achieved. The programme leader should maintain a progress register, supported by a Gantt chart, defining the change, expected

Table 9.3 Derived importance of projects

Projects	Cost take out High: 10 Low: 1	Productivity High: 5 Low: 1	Cost avoidance High: 2 Low: 0	Employee satisfaction High: 3 Low: 1	Current performance Low: 10 High: 1	Feasibility High: 10 Low: 1	Derived importance Max. score: 40
Design packaging	9	2	0	2	7	8	28
Improve CEE	5	4	2	2	10	10	33
Passes control	1	2	2	1	5	9	20
New product development	7	1	2	3	2	2	17

benefits, resources, timescale and expenditure (to budget), and show people responsible for actions.

This phase of review and feedback involves a continuous need to sustain what has been achieved and to identify further opportunities for improvement. It is good practice to set fixed dates for review meetings as follows:

- Milestone review (at least quarterly) – Executive Steering Committee, torchbearer and programme leader
- Programme review (monthly) – programme leader and team, with a short report to torchbearer.

The problems/hold-ups experienced during projects are identified during the programme review, with the aim of the project team taking action to resolve sticking points; if necessary, requests are made of the Executive Steering Committee for additional resources.

Step 7: Measurement of success

The fundamental characteristic of a Six Sigma or FIT SIGMA programme that differentiates it from a traditional quality programme is that it is results orientated. Effective measurement is the key to understanding the operation of the process, and this forms the basis of all analysis and improvement work. In a construction project the milestones are tangible – they are physically obvious – but in a change programme such as FIT SIGMA the changes are not always apparent. It is important to measure, display and celebrate the achievement of milestones in a FIT SIGMA programme. In Chapter 6 we emphasized the importance of performance management to improve and sustain the results of a FIT SIGMA programme. The process and culture of measurement must start during the implementation of changes.

Our experience is that the components of measurement of success should include:

- Project tracking
- FIT SIGMA metrics
- Balanced scorecard
- Self-assessment review (Baldridge or EFQM).

There are software tools available such as Minitab (www.minitab.com) and KISS (Keep It Simple Statistically; www.airacad.com/transact.html) for detailed tracking of larger Six Sigma projects. However, in most programmes the progress of savings generated by each project can be monitored on an Excel spreadsheet. We recommend that summaries of results are reported and displayed each month. Examples of forms of displays are shown in Figures 9.6 and 9.7.

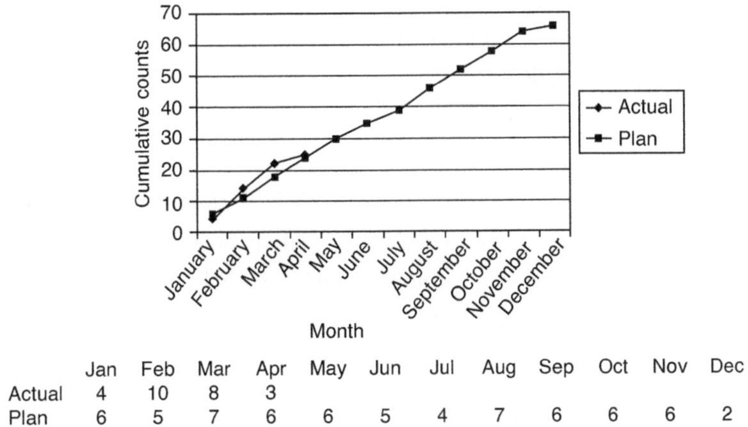

	Jan	Feb	Mar	Apr	May	Jun	Jul	Aug	Sep	Oct	Nov	Dec
Actual	4	10	8	3								
Plan	6	5	7	6	6	5	4	7	6	6	6	2

Figure 9.6 Projects planned and completed 2002.

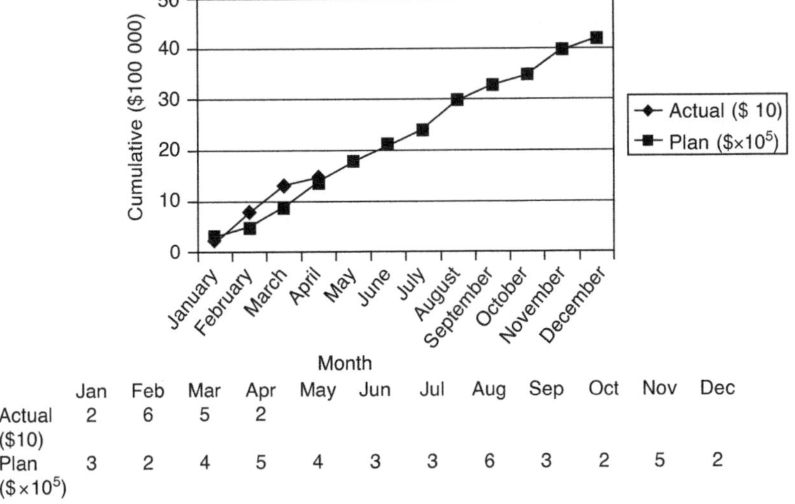

	Jan	Feb	Mar	Apr	May	Jun	Jul	Aug	Sep	Oct	Nov	Dec
Actual ($10)	2	6	5	2								
Plan ($ $\times 10^5$)	3	2	4	5	4	3	3	6	3	2	5	2

Figure 9.7 Value of planned and completed projects.

FIT SIGMA metrics

FIT SIGMA metrics are required to analyse the reduction in process variance and the reduction in the rate of defects resulting from the appropriate tools and methodology. A word of caution: Black Belts can get caught up with the elegance of statistical methods and develop a statistical cult. Extensive use of variance analysis is not recommended.

The following FIT SIGMA metrics are useful, easy to understand, and easy to apply:

- Cost of Poor Quality (COPQ)
- Defects per Million Opportunities (DPMO)
- First Pass Yield (FPY).

$$\text{COPQ} = \frac{\text{Internal failure \$ + External failure \$} + \text{Appraisal and prevention \$ + Lost opportunity \$}}{\text{Monthly sales \$}}$$

$$\text{DPMO} = \frac{\text{Total number of defects} \times 1\,000\,000}{\text{Total units and opportunities per unit}}$$

$$\text{FPY} = \frac{\text{Number of units completed without defects and rework}}{\text{Number of units started}}$$

From measuring and monitoring FIT SIGMA metrics each month, opportunities for further improvement will be identified.

As we have already emphasized, a carefully designed Balanced Scorecard (see Chapter 6) is essential for improving and sustaining business performance. It is generally viewed that the Balanced Scorecard is applicable for a stable process and thus should be appropriate after the completion of the FIT SIGMA programme. This may be so, but unless the measures of the Balanced Scorecard are properly defined and designed for the purpose at an early stage, its effectiveness will be limited. Therefore, we strongly recommend that during the FIT SIGMA programme the basics of the Balanced Scorecard to manage the company-wide performance system should be established (see Figures 6.8 and 6.9).

The fourth component of measurement is the 'self-assessment and review' process. There are two options to monitor the progress of the business resulting from the FIT SIGMA programme:

1. A simple checklist to assess the overall progress of the programme, or
2. A proven self-assessment process such as the European Foundation of Quality Management (EFQM) or the American Malcolm Baldridge system.

We recommend the second option. In the initial health-check appraisal stage we recommended using EFQM or Baldridge, and thus the methodology will have already been applied. Additionally it gives further experience in the self-assessment process, which will enable future sustainability. Finally, it will provide the foundation should the organization wish at a later stage to apply for an EFQM or Baldridge award.

Step 8: Culture change

A culture change must *not* begin by replacing middle management with imported 'Black Belts'. Winning over, not losing, middle management is essential to the success of FIT SIGMA – or for that matter any quality initiative.

What is required is that the all-important middle management, and everyone else in the organization, understands what FIT SIGMA is, and has the culture of quality. An understanding of Deming's fourteen points (see Chapter 2) would be a sound start.

The FIT SIGMA Culture is:

1. Total vision and commitment of top management throughout the programme
2. Emphasis on measured results and the rigour of project management
3. Focus on training with short-term projects and results, and long-term people development
4. Use of simple and practical tools
5. A total approach across the whole organization (holistic)
6. Leveraging results by sharing best practice with business partners (suppliers and customers)
7. Sustaining improvement by knowledge management, regular self-assessment and senior management reviews.

Air Academy www.airacad.com (April, 2002) claim that it is important to understand the culture type of an organization to ensure the culture change necessary for the success of a Six Sigma programme. The culture types they identify are:

1. *Clan culture.* The organization is considered to be people-orientated. It is a nice place to work where people share similar interests, much like a country club – i.e. they are clan-like, they all have similar beliefs and values.
2. *Hierarchical culture.* The organization has a formalized top-down structure, and people are governed by rules and procedures.
3. *Enterprise culture.* The organization is goal-orientated. Results are measured and members are competitive. People are primarily concerned with getting the job done.
4. *Adhocracy culture.* The organization is dynamic, creative and entrepreneurial. People are proactive and take risks, are innovative and look for alternatives.

FIT SIGMA requires a balanced culture comprising key characteristics of the above four types. If an organization is predominantly of one type, some cultural change will be required. Training deployment (see Step 5 of Figure

9.3) includes training for culture change, but training alone will not transform the mindset required for FIT SIGMA.

Creating a receptive culture

An often-asked question is, how do we change culture?

It all begins with Vision. The vision of quality must begin with the Chief Executive. If the Chief Executive has a passion for quality and continuous improvement, and if this passion can be transmitted down through the organization, then paradoxically the ongoing driving force will be from the bottom up rather than being enforced from above, and with everyone sharing the same vision. For similar viewpoints *re*. TQM see Crosby (1979), Ishikawa (1985), Schonberger (1986), Albrecht (1988), Collins and Porras (1991), Creech (1994), Dulewicz *et al.* (1995) and Gabor (2000).

The word 'vision' suggests an almost mystical occurrence (Joan of Arc) or an ideal (such as expressed by Martin Luther King, 'I have a dream . . .'). The same connotation is found when looking at vision in the organizational context; a leader with a vision is a leader with a passion for an ideal. However, '. . . unless the vision can happen, it will be nothing more than a dream' (Wright, 1996, p. 20; see also El-Namki, 1992 and Langeler, 1992). To make a vision happen within an organization, there has to be a cultural fit. Corporate culture is the amalgam of existing beliefs, norms and values of the individuals who make up the organization – 'the way we do things around here' (Peters and Waterman, 1982; Peters and Austin, 1986). The leader may be the one who articulates the vision and makes it legitimate, but unless it mirrors the goals and aspirations of the members of the organization at all levels the vision won't happen (Albrecht, 1988). As Stacey (1993, p. 234) says, 'the ultimate test of a vision is if it happens'.

Culture and values are deep-seated and may not always be obvious to members. As well as the seemingly normal aversion to change by individuals, often there is a vested interest for members of an organization to resist change. Middle management is often more likely to resist change than are other members. Machiavelli (1513) wrote: 'It must be considered that there is nothing more difficult to carry out, nor more doubtful to success, nor more dangerous to handle, than to initiate a new order of things'. Human nature hasn't really changed much since the sixteenth century!

Organizations are made up of many individuals, each with his or her own set of values. The culture of an organization is how people react or do things when confronted with the need to make a decision. If the organization has a strong culture, each individual will instinctively know how things are done and what is expected. Conversely, if the corporate culture is weak, the individual may not react in the manner that management would hope (Peters and Waterman, 1982; Peters and Austin, 1986; Carnall, 1999).

To engineer or change a culture, there has to be leadership from the top. Leading by example might seem to be a cliché, but unless the Chief Executive can clearly communicate and demonstrate by example a clear policy, how will the rest of the people know what is expected? (Foreman and Money, unpublished work; Foreman, 1996, 2000). Leadership does not have to be charismatic, but it has to be honest. Leadership does not rely on power and control. Basu and Wright (1998) find that real leaders communicate face-to-face, and not by memos.

Mission statement to signal change

A new mission statement would seem to be a logical way for a Chief Executive to signal a change in direction for an organization. Ideally the mission statement should be a true statement as to the reason for being of the organization. It should be realistic, and state the obvious. Profit is not a dirty word (Friedman, 1970), and customer service is important (Zeithaml *et al.*, 1990; Kotler, 1999). Generally, the key resource lacking in many organizations is quality people (Barlett and Ghoshal, 1994; Mintzberg, 1996; Knuckey *et al.*, 1999). Therefore it would seem obvious for any mission to say we are in business to make a profit, and we will make a profit by providing the customers with what they want, and that we recognize that our most important resource in making this mission happen is our people. It is important that the new mission is in tune with what the people of the organization believe (the culture), and to achieve a mission that fits the culture it would seem sensible to get the involvement and interest of all the staff in writing the new mission. Thus in this manner a change in culture could begin with the determination and the buy-in by staff into the new mission.

Learning for change

> *If employees, organization wide, are going to accept change, and themselves individually change, they will need to learn certain skills. Skills such as;*
> *understanding work processes,*
> *solving problems,*
> *making decisions, and*
> *working with others in a positive way.*
> *All these types of skills can be taught. The main message that has to be learnt is the need for cultural change, and for people to trust each other. In particular management has to win the trust of lower level staff and have to learn how to change from autocratic management to coaching and mentoring. Lower level staff, in turn, have to learn to trust management.*

(Wright, 1999, p. 219; see also Hall, 1999 and Axelrod, 2001, who express similar views.) Once this has been achieved the culture will be such that the organization will be in tune with the philosophy of FIT SIGMA, and the ninth step of FIT SIGMA (improve and sustain) will be second nature.

Communication

Finally, the key to sustaining a FIT SIGMA culture is communication. Methods of communication include:

- An intranet FIT SIGMA website, specifically developed, or clearly visible in the corporate web site
- Specially produced videos or CDs
- A FIT SIGMA monthly newsletter
- Internal emails, voicemails and memos with updated key messages – *not* slogans such as work smarter not harder and other tired clichés
- Milestone celebrations
- Staff get-togethers, such as special morning teas, Friday afternoon social hour, 'town hall' type meetings
- An 'ideas factory' or 'think tank' to encourage suggestions and involvement from employees.

Step 9: Improve and sustain

'Improve and sustain' is the cornerstone of a FIT SIGMA programme. This step is similar to the fifth stage of team dynamics for project teams (Forming, Storming, Norming, Performing and Mourning). In the Mourning stage the project team disbands and members move onto other projects or activities. They typically regret the end of the project and the break-up of the team, and the effectiveness or maintenance of the new method and results gradually diminish. Chapter 6 discussed in some detail the fact that to achieve sustainability the following four key processes must be in place:

1. Performance management
2. Senior management review
3. Self-assessment and certification
4. Knowledge management.

The 'end game' scenario should be carefully developed long before the end of the programme. There may not be a sharp cut-off point like a project handover, and the success of the scenario is in the making of a smooth transition without disruption to the ongoing operation of the business.

As part of the performance management the improvement targets should be gradually and continuously stretched, and more advanced tools considered for introduction. For example, the DFSS (Design for Six Sigma) is resource hungry, and can be considered at a later stage in a FIT SIGMA programme. With Six Sigma the aim is to satisfy customers with robust 'zero defect' manufactured products, and to do so DFSS is fully deployed, covering all elements of manufacturing, design, marketing, finance, human resource, suppliers and key customers (including supplier's suppliers, and customer's customers).

At an advanced stage of the programme, milestone reviews should be included in senior management operational review team meetings (such as the sales review meetings and operational planning meetings/committees), i.e. not only the FIT SIGMA Executive Steering Committee.

We recommend that a pure play EFQM (or other form of self-assessment) be incorporated as a six-monthly feature of the FIT SIGMA programme. Even if the company gains an EFQM or Baldridge award, the process MUST continue indefinitely.

Two specific features of knowledge management need to be emphasized. First, it is essential that the company seeks leverage from FIT SIGMA results by rolling it out to other business units and main suppliers. Second, it is equally important to ensure that career development and reward schemes are firmly in place to retain the highly trained and motivated Black Belts. The success of sustainability of FIT SIGMA is when the culture becomes simply 'this is the way we do things'.

Time scale

The time scale of FIT SIGMA implementation will last several months, and is of course variable. The time not only depends on the nature or size of the organization, but also on the business environment and the resources available. Four factors can favourably affect the time scale:

1. Full commitment of top management and the Board
2. Sound financial position
3. Correct culture (workforce receptive to change)
4. A competitive niche in the marketplace.

It is good practice to prepare a Gantt chart containing the key stages of the programme, and to monitor the progress. Figure 9.8 shows a typical timetable for a FIT SIGMA programme in a single-site medium-sized company. The diagram shows an order of magnitude only, and the sequence could well vary. The timeline is not linear, stages overlap, and frequent looking back to learn for future progress should occur.

FIT SIGMA for stalled Six Sigma

Some organizations have already attempted to implement a Six Sigma (or a TQM) programme, but it has stalled. Results are not being achieved and enthusiasm is waning; in some cases the programme has effectively been abandoned. The reasons for stalling are various, but often include an economic downturn (such as experienced in the telecommunications industry in 2001), a change in top management, or a merger or takeover.

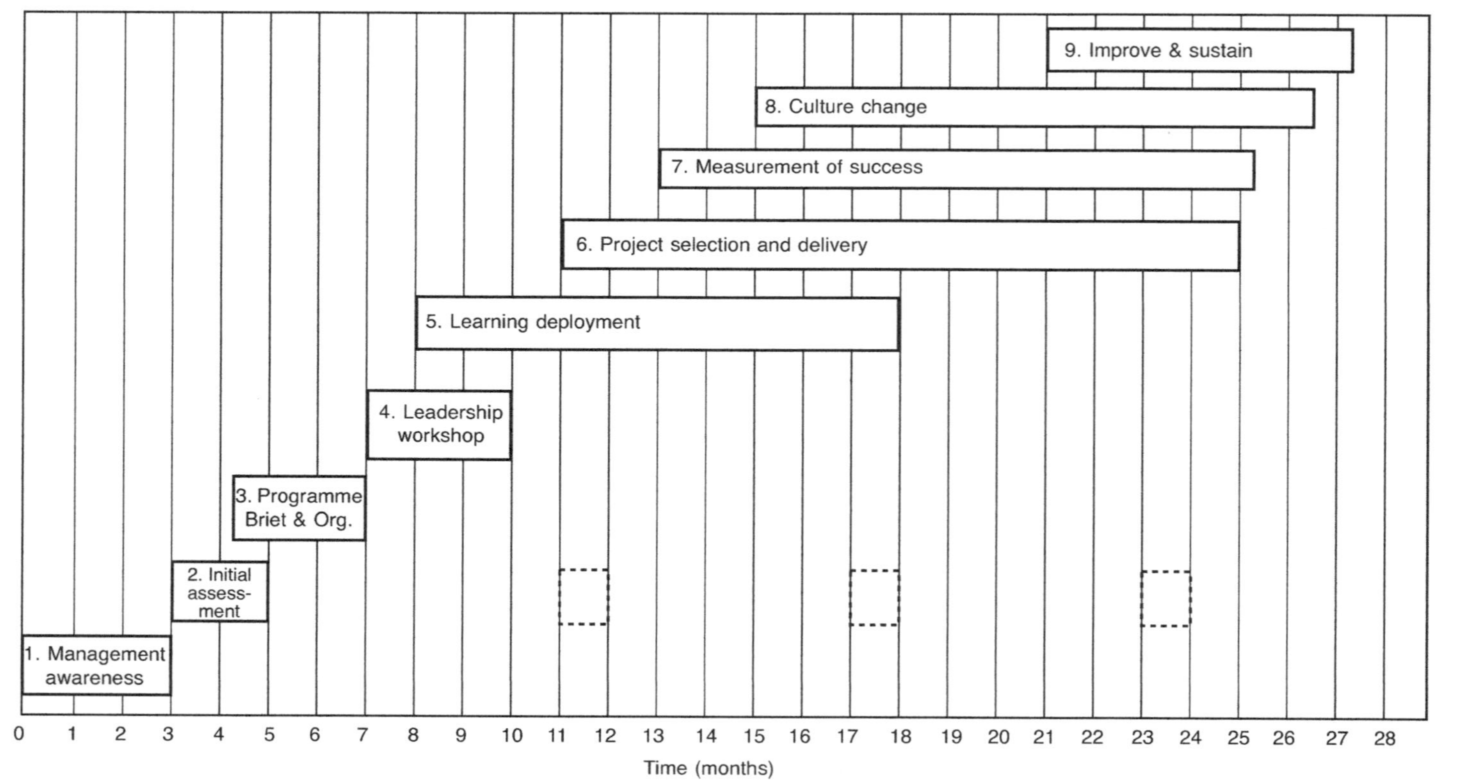

Figure 9.8 A typical timetable for a FIT SIGMA™ programme.

FIT SIGMA, not a quick fix

During restructuring, or if the company is in survival mode, the implementation of FIT SIGMA is not appropriate. FIT SIGMA is not a quick fix. After the short-term cost saving measures of a survival strategy, when the business has stabilized and a new management team is in place, then Six Sigma can be restarted – but this time done correctly, using the FIT SIGMA approach.

It is likely when restarting that many of the steps, including training/learning deployment, will not need to be repeated. However, in a restart there is one big issue that makes life more difficult, and that is credibility. How do you convince all the people that it will work the second time around? This will put special pressure on Step 8, culture change. The employees could well be tired of excessive statistics and complex Six Sigma tools. Selection of appropriate tools is a strong feature of FIT SIGMA.

The FIT SIGMA programme for a re-starter will naturally vary according to the condition of the organization, but the programme can be adapted within the framework shown in Figure 9.3. The guidelines for each step are:

1. Management awareness. If this is not present, the programme cannot re-start
2. Initial assessment. This has to be the re-start point – where are we, where do we want to go?
3. Programme brief and organization. The programme will need to be re-scoped and new teams formed.
4. Leadership workshop. This will be essential, even if management has not changed.
5. Training/learning deployment. Appropriate tools should be selected. If past team members (in particular Black Belts) are not happy with terminology of the old programme, new terms should be used. The title 'Black Belt' in itself is not sacrosanct and might be changed. If the 'old' experts are still in the organization, then training time might be reduced – even a one-week workshop might be sufficient.
6. Project selection and delivery. This is the same as for the full FIT SIGMA programme – harvest the hanging fruit.
7. Measurement of success. Review the old measures – what worked and what didn't – and follow the full FIT SIGMA programme.
8. Culture change. This is critical. Top management support must be extremely obvious, and reward and appraisal systems aligned to FIT SIGMA.
9. Improve and sustain. This is the same as for FIT SIGMA.

FIT SIGMA for successful companies

William Stavropoulos, the CEO of Dow Chemical, is reported to have once said: 'the most difficult thing to do is to change a successful company'. It is true that employees of companies enjoying a high profit margin with some dominance in the market are likely to be complacent and to feel comfortable with the *status quo*. Perhaps it is even more difficult to stay at the top or sustain success if the strategy and processes are not adaptable to change – as Darwin (*c*. 1860) said: 'It is not the strongest species that survive, nor the most intelligent, but the ones most responsive to change'. It is possible for management of some companies, after the completion of a highly successful Six Sigma programme, to find their attention diverted to another major initiative such as e-Business or Business-to-Business Alliances. Certainly new initiatives must be pursued, but at the same time the long-term benefits that could be achieved from Six Sigma should not be lost. FIT SIGMA for sustainability, staying healthy, is the answer.

If a company has succeeded with Six Sigma, then the time is now right to move onto FIT SIGMA to achieve Step 9 – Improve and sustain.

External consultants

Many companies, especially SMEs, are often concerned with the cost of consultants for a Six Sigma programme. Large consulting firms and academies for Six Sigma could well expect high front-end fees. With FIT SIGMA, our approach is to be selective in the use of outside consultants – that is, use outside consultants to train your own experts, and to supplement your own expertise and resource when necessary. No consultant will know you own company as well as your own people will. For this reason we do not favour the use of an external consultant as a programme leader.

In a FIT SIGMA programme, the best use of consultants is in:

- Step 2 (Initial assessment). Here a Six Sigma expert or an EFQM consultant may be used to train and guide your team.
- Steps 4 and 5 (Leadership workshop, and Training/learning deployment). Outside consultants will be needed to facilitate the leadership workshops, and to train your own Black Belts. Once trained, your own Black Belts will train Green Belts and develop new Black Belts.
- Step 8 (Culture change). An outside consultant is best suited to develop a change management plan for change of culture.

Summary

This chapter provides practical guidelines for making it all happen. Many a Six Sigma exercise started with high expectations and looked good on paper. Many an organization has been impressed by success stories of Six Sigma, but is unsure how to start.

The implementation plan shown here will enable *any* organization at any stage of a Six Sigma initiative to follow a proven path to success and to sustain benefits. Our implementation plan has nine steps, beginning with

Management awareness and leading to the *ongoing* step of Improve and Sustain. There is no end!

In the spirit of FIT SIGMA, fit for purpose, this framework can be adjusted and customized to the specific needs of any organization.

At all stages of the programme it is essential that not only are the Executive Steering Committee and the torchbearer kept informed (the torchbearer will keep the Board informed), but also that there is open communication with all members of the organization, so that everyone is aware of the aims, activities and successes of the programme.

References

Chapter 1

Basu, R. and Wright, J.N. (1998). *Total Manufacturing Solutions*. Oxford, Butterworth-Heinemann.
Deming, W.E. (1986). *Out of the Crisis*. In Wright, J.N. (1999). *The Management of Service Operations*. London, Cassell.
Erwin, J. and Douglas, P.C. (2000). *It's not difficult to change company culture*. Supervision; Burlington; 61(11).
Turner, J.R. (1999). *The handbook of Project Based Management: improving the processes for achieving Strategic Objectives*. 2nd Edition. London, McGraw-Hill.
Womack, J., Jones, D. and Roos, D. (1990). *The Machine that Changed the World*. New York, Rawson and Associates.

Chapter 2

Basu, R. and Wright, J.N. (1998). *Total Manufacturing Solutions*. Oxford, Butterworth-Heinemann.
Carlzon, J. (1989). *Moments of Truth*. New York, Harper Row.
Crosby, P.B. (1979). *Quality without Tears*. In Turner, J.R. (1993). *The Handbook of Project Based Management: Improving the processes for achieving strategic objectives*. London, McGraw-Hill.
Dale, B.G. (1999). (Ed.) *Managing Quality*. New York, Prentice Hall.
Deming, W.E. (1986). *Out of the Crisis*. In Wright, J.N. (1999). *The Management of Service Operations*. London, Cassell.
Feigenbaum, A.V. (1983). *Total Quality Control*. New York, McGraw-Hill.
Ferguson, I. and Dale, B.G. (1999). In Dale, B.G. (1999). (Ed.) *Quality Function Deployment*. New York, Prentice Hall.
Fry, T.D., Steele, D.C. and Sladin, B.A. (1994). A Service Orientated Manufacturing Strategy. *International Journal of Operations and Production Management*, Vol. 14(10). pp. 17–29.
Gabor, A. (2000). *He made America think about Quality*. Fortune, October 142(10). New York.
Harrison, A. (1998). Manufacturing Strategy and the concept of world class manufacturing. *International Journal of Operations and Production Management*, Vol. 18(4) pp. 397–408).
Hayes, R.H. and Wheelwright, S.C. (1984). *Restoring our Competitive Edge: Competing through Manufacturing*. John Wiley and Sons, New York.
Imai, M. (1986). *Kaizen: the key to Japan's competitive success*. New York, Random House Business Division.
Ishikawa, K. (1979). *Guide to Quality Control*. Tokyo, Asian Productivity Organisation.

Ishikawa, K. (1985). *What is Total Quality Control? The Japanese Way*, trans. D.J. Lu, Englewood Cliffs, Prentice Hall.

Juran, J.M. (1988). *Juran on Planning for Quality*. New York, Free Press.

Juran, J.M. (1989). *Juran on Leadership for Quality: an executive handbook*. New York, Free Press.

Knuckey, S., Leung-Wai, J. and Meskill, M. (1999). *Gearing Up: A study of Best Manufacturing Practice in New Zealand*. Wellington, Ministry of Commerce.

Oakland, J.S. (2000). *Total Quality Management: Text with Cases*. (2nd Edition). Oxford, Butterworth-Heinemann.

Pyzdek, T. (2000). *The Six Sigma Revolution*. HYPERLINK "http:/www.pyzdek.com" www.pyzdek.com/six-sigma-revolution.htm, 2 May 2002.

Sayle, A.J. (1991). *Meeting ISO 9000 In a TQM World*. Great Britain, AJSL.

Schonberger, R. (1986). *World Class Manufacturing*. New York, Free Press.

Shingo, S. (1985). *A Revolution in Manufacturing: The SMED System*. Cambridge MA, Productivity Press.

Taylor, F.W. (1987). In Wright, J.N. (1999). *The Management of Service Operations*. London, Cassell.)

Walton, M. (1986). *The Deming Method*. New York, Perigree Books.

Womack, J., Jones, D. and Roos, D. (1990). *The Machine that Changed the World*. New York, Rawson and Associates.

Chapter 3

Fisher, R.A. (1925). *Statistical Methods for Research Workers*. Edinburgh, Oliver and Boyd.

Fry, T.D., Steele, D.C. and Sladin, B.A. (1994). A Service Orientated Manufacturing Strategy. *International Journal of Operations and Production Management* 14(10), 17–29.

Hayes, R.H. and Wheelwright, S.C. (1984). *Restoring our Competitive Edge: Competing Through Manufacturing*. John Wiley and Sons, New York.

Turner, J.R. (1993). *The Handbook of Project Based Management: Improving the processes for achieving strategic objectives*. London, McGraw-Hill.

Wild, R. (2002). *Operations Management*. London, Continuum.

Chapter 5

Walker, C. and Guest, R. (1952). *The Man on the Assembly Line*. Cambridge, Harvard University Press.

Womack, J., Jones, D. and Roos, D. (1990). *The Machine that Changed the World*. New York, Rawson and Associates.

Chapter 6

Basu, R. and Wright, J.N. (1998). *Total Manufacturing Solutions*. Oxford. Butterworth-Heinemann.

Deming, W.E. (1982) *Quality, Productivity and Competitive Position*. Cambridge, Mass: MIT Centre for Advanced Learning.

Deming, W.E. (1986). *Out of the Crisis*. In Wright, J.N. (1999). *The Management of Service Operations*. London, Cassell.

Drucker, P. (1995). *US News and World Report.* In Schmidt, S.R., Kiemele, M.J. and Berdine, R.J. (1999). *Knowledge Based Management.* Air Academy, pp. 17.

Juran, J.M. (1989). *Juran on Leadership for Quality: an executive handbook.* New York, Free Press.

Kaplan, R.S. and Norton, D.P. (1996). Using the Balanced Scorecard as a Strategic Management System. *Harvard Business Review*, January–February, pp. 75–85.

Oakland, J.S. (2000). *Total Quality Management: Text with Cases.* (2nd Edition). Oxford, Butterworth-Heinemann.

Shewart, W.A. (1931). *Economic Control of Quality of Manufactured Product.* New York, Van Nostrand Co.

Welch, J. and Byrne, J.A. (2001). *Jack: Straight from the Gut.* Warner Books.

Wild, R. (2002). *Operations Management.* London, Continuum.

Chapter 7

Basu, R. and Wright, J.N. (1998). *Total Manufacturing Solutions.* Oxford. Butterworth-Heinemann.

Berry, L.L., Parasuraman, A. and Zeithaml, V.A. (1988). *The Service Quality Puzzle.* In Wright, J.N. (1999). *The Management of Service Operations.* London, Cassell.

Christopher, M. (1992). *Logistics and Supply Chain Management.* Pitman Publishing.

Lewis, B.R. (1994). In Dale, B.G. *Managing Quality.* Prentice Hall.

Parasuraman, A., Zeithaml, V.A. and Berry, L.L. (1985). A Conceptual Model of Service Quality and its Implications for Future Research. *Journal of Marketing*, 49, Fall, pp 41–50.

Parasuraman, A., Zeithaml, V.A. and Berry, L.L. (1991). Understanding Customer Expectations of Service. *Sloan Management Review*, 32(3), pp. 39–48.

Wild, R. (2002). *Operations Management.* London, Continuum.

Zeithaml, V.A., Parasuraman, A. and Berry, L.L. (1990). *Delivering Quality Service: Balancing Customer Perceptions and Expectations.* New York, The Free Press.

Chapter 8

Goldratt, E.M. (1997). *Critical Chain.* Great Barrington MA, The North River Press.

Michael, N. and Burton, C. (1991). *Basic Project Management.* Auckland, Reed Books.

Obeng, E. (1994). *All Change! The Project Leaders' Secret Handbook.* London, Pitman.

Turner, J.R. (1993). *The Handbook of Project Based Management: Improving the processes for achieving strategic objectives.* London, McGraw-Hill.

Turner, J.R. (1999). *The Handbook of Project Based Management: Improving the processes for achieving Strategic Objectives.* 2nd Edition. London, McGraw-Hill.

Wilemon, D.L. and Baker, B.N. (1983). *Some Major Research Findings Regarding the Human Element in Project Management.* In Cleland, D.I. and King, R.W. (Eds) *Project Management Handbook.* New York, Van Nostrand Reinhold Co.

Chapter 9

Albrecht, K. (1988). *At America's Service.* Homewood, Il. Dow Jones-Irwin.

Axelrod, R.H. (2001). 'Changing the way we change organisations', *The Journal of Quality and Participation*, Spring, 24(1), pp. 22–27.

Barlett, C. and Ghoshal, S. (1994). *The Changing Role of Top Management. INSEAD working papers*, London Business School Library.

Basu, R. and Wright, J.N. (1998). *Total Manufacturing Solutions.* Oxford. Butterworth-Heinemann.

Carnall, C. (1999). *Managing Change in Organisations.* London, Prentice Hall Europe.

CCTA (1999). *Central Computer and Telecommunication Agency. Managing Successful Programmes.* London, Stationery Office.

Collins, J.C. and Porras, J.I. (1991). Organizational Vision and Visionary Organizations. *Californian Management Review*, 34, pp. 30–52.

Creech, B. (1994). *The Five Pillars of TQM.* New York, Truman Talley Books.

Crosby, P.B. (1979). *Quality without Tears.* In Turner, J.R. (1993). *The Handbook of Project Based Management: Improving the processes for achieving strategic objectives.* London, McGraw-Hill.

Darwin, C. (1998). *The Origin of Species.* (Edited by Suriano, G.) Grammercy Publishers.

Dulewicz, V., MacMillan, K. and Herbert, P. (1995). Appraising and Developing the Effectiveness of Boards and their Directors. *Journal of General Management*, 20(3) pp. 1–19.

El-Namki, M.S.S. (1992). Creating A Corporate Vision. *Long Range Planning*, 25(6) pp. 25–29.

Foreman, S.K. and Money, A.H. (1995). *Internal Marketing: Concepts, Measurement and Application.* Henley Management College: Oxon, Unpublished Paper.

Foreman, S.K. (1996). *Internal Marketing.* In Turner, J.R., Grude, K.V. and Thurloway, L. (Eds), *The Project Manager as Change Agent.* London, McGraw-Hill.

Foreman, S.K. (2000). *Internal Marketing.* In Lewis, B. and Varey, R. (Eds), *Internal Marketing*, Psychology Press.

Friedman, M. (1970). A Friedman Doctrine: The Social Responsibility of Business is to Increase its Profits. *New York Times Magazine*, September 13, 32.

Gabor, A. (2000). *He Made America think about Quality.* Fortune, October 142(10). New York.

Hall, J. (1999). Six Principles for Successful Business Change Management. *Management Services*, 43(4), pp. 16–18.

Hammer, M. and Champy, J. (1993). *Re-Engineering the Corporation.* London, Nicholas Brealey Publishing.

Ishikawa, K. (1985). *What is Total Quality Control? The Japanese Way*, trans. D.J. Lu, Englewood Cliffs, Prentice Hall.

KISS: *Keep It Simple Statistically.* www.airacad.com/transact.html

Knuckey, S., Leung-Wai, J. and Meskill, M. (1999). *Gearing Up: A Study of Best Manufacturing Practice in New Zealand.* Wellington, Ministry of Commerce.

Kotler, P. (1999). *Kotler on Marketing.* New York, Free Press.

Langeler, G.H. (1992). The Vision Trap. *Harvard Business Review*, March–April, pp. 46–49.

Machiavelli, N. (1513). *The Prince.* Translated by Luigi Ricci, revised by Vincent, E.R.P. (1952). New York: New American Library of World Literature.

Mintzberg, H. (1996). Musings on Management. *Harvard Business Review*, July–August, pp. 61–67.

Peters, T. and Waterman, J.R. (1982). *In Search of Excellence.* New York, Harper Row.

Peters, T. and Austin, N. (1986). *A Passion for Excellence.* London, Fortune.

Schonberger, R. (1986). *World Class Manufacturing.* New York, Free Press.

Stacey, R.D. (1993). *Strategic Management and Organizational Dynamics.* London, Pitman Publishing.

Turner, J.R., Grude, K.V. and Thurloway, L. (1996). *The Project Manager as Change Agent.* London, McGraw-Hill.

Wright, J.N. (1996). Creating a Quality Culture. *International Journal of General Management*, 21(3), pp. 19–29.

Wright, J.N. (1999). *The Management of Service Operations.* London, Cassell.

Zeithaml, V.A., Parasuraman, A. and Berry, L.L. (1990). *Delivering Quality Service: Balancing Customer Perceptions and Expectations.* New York, The Free Press.

Glossary

Best practice Refers to any organization that performs as well or better than the competition in quality, timeliness, flexibility, and innovation. Best practice should lead to world-class performance.

Black Belts Experts in Six Sigma methods and tools. Tools include statistical analysis. Black Belts are project leaders for Six Sigma initiatives; they also train other staff members in Six Sigma techniques.

BPR (Business Process Re-engineering) This has been described as a manifesto for revolution. The approach is similar to taking a clean piece of paper and starting all over by identifying what is really needed to make the mission of the organization happen.

Brainstorming A free-wheeling group session for generating ideas. Typically a group meeting of about seven people will be presented with a problem, and each member will be encouraged to make suggestions without fear of criticism. One suggestion will lead to another. All suggestions, no matter how seemingly fanciful, are recorded, and subsequently analysed. Brainstorming is useful for generating ideas for further detailed analysis.

Cause and effect diagram The cause and effect, fishbone or Ishikawa diagram was developed by Kaoru Ishikawa. The premise is that generally when a problem occurs the effect is very obvious, and the temptation is to treat the effect. With the Ishikawa approach the causes of the effect are sought. Once these are known and eliminated, the effect will not be seen again. For example, working overtime is an effect; adding extra staff does not remove the cause. The question is, what caused the situation that led to overtime being worked?

COPQ (Cost of Poor Quality) The cost of poor quality is made up of costs arising from internal failures, external failures, appraisal, prevention and lost opportunity costs – in other words, all the costs that arise from non-conformance to a standard. Chapter 3 discusses COPQ in some detail.

CTQ (Critical to Quality) Six Sigma refers to CTQS. This simply means the identification of factors that are critical for the achievement of a level of quality.

DFFS (Design for FIT SIGMA™) The steps are Define, Measure, Analyse, Design and Validate (see Chapter 3 for detailed discussion).

DMAIC The cycle of Define, Measure, Analyse, Improve and Control.

DOE The process of examining options in the design of a product or service. Controlled changes of input factors are made and the resulting changes to outputs noted. Losses from poor design include not only direct loss to the company from reworking and scrap, but also include those owing to user downtime due to equipment failure, poor performance and unreliability. Poor customer satisfaction will lead to further losses by the company as market share falls.

DPMO (Defects per million opportunities) The basic measure of Six Sigma. It is the number of defects per unit divided by the number of opportunities for defects multiplied by 1 000 000. This number can be converted into a Sigma value – for example, Six Sigma = 3.4 per million opportunities.

E-business Electronic-business is more than the transfer of information using information technology; it is the complex mix of processes, applications and organizational structures.

EFQM (European Foundation for Quality Management) Derived from the American Malcolm Baldridge Quality Award, this is an award for organizations that achieve world-class performance as judged by independent auditors against a checklist. The checklist is detailed and extensive, and covers Leadership, People Management, Policy and Strategy, Partnerships and Resource, Processes, People Satisfaction, Customer Satisfaction, Impact on Society, and Business Results.

Fishbone diagram The fishbone, Ishikawa, or cause and effect diagram was developed by Kaoru Ishikawa. The premise is that generally when a problem occurs the effect is very obvious, and the temptation is to treat the effect. With the Ishikawa approach the causes of the effect are sought. Once these are known and eliminated, the effect will not be seen again. For example, working overtime is an effect; adding extra staff does not remove the cause. The question is, what caused the situation that led to overtime being worked?

FIT SIGMA™ (see also **TQM, Six Sigma** and **Lean Sigma**) FIT SIGMA incorporates all the advantages and tools of TQM, Six Sigma and Lean Sigma. The aim is to get an organization healthy (fit) by using appropriate tools for the size and nature of the business (fitness for purpose) and to sustain a level of fitness. FIT SIGMA is a holistic approach.

Flow process chart A flow process chart sets out the sequence of the flow of a product or a procedure by recording all the activities in a process. The chart can be used to identify steps in the process, value-adding activities and non-value-adding activities.

FMEA (Failure Mode and Effect Analysis) This was developed in the aerospace and defence industries. It is a systematic and analytical quality planning tool for identifying, at the design stage of new products or services, what could go wrong during manufacture, or when in use by the customer. It is an iterative process, and the points examined are:

- What the function is
- Potential failure modes
- The effect of potential failure
- Review of current controls
- Determination of risk priority (occurrence, detection, and severity of failure)
- Identification of corrective actions to eliminate failures
- Monitoring of corrective actions and countermeasures.

FPY (First Pass Yield) Also known as RTY, this is the ratio of the number of completely defect-free units (without any kind of rework during the process) at the end of a process and the total number of units at the start of a process. The theoretical throughput rate is often regarded as the number of units at the start of the process. RTY/FPY is used as a key performance indicator to measure overall process effectiveness.

Green Belts Staff trained to be Six Sigma project leaders, Green Belts work under the guidance of Black Belts (see Black Belts).

Histogram A histogram is a descriptive and easy to understand chart of the frequency of occurrences. It is a vertical bar chart with the height of each bar representing the frequency of an occurrence.

Input–process–output diagram All operations or processes have inputs and outputs, and the process is the conversion of inputs into outputs. Analysis of inputs should be made to determine factors that influence the process – for example, input materials from suppliers meeting specification, delivery on time and so on. Examples of input–process–output diagrams for service and manufacturing industries are shown in Figures 3.5 and 3.6.

Ishikawa The Ishikawa, fishbone or cause and effect diagram was developed by Kaoru Ishikawa. The premise is that generally when a problem occurs the effect is very obvious, and the temptation is to treat the effect. With the Ishikawa approach the causes of the effect are sought. Once these are known and eliminated, the effect will not be seen again. For example, working overtime is an effect; adding extra staff does not remove the cause. The question is, what caused the situation that led to overtime being worked?

ISO 9000 To gain ISO 9000 accreditation, an organization has to demonstrate to an accredited auditor that they have a well-documented standard and consistent process in place that achieves a defined level of quality or performance. ISO accreditation will give a customer confidence that the product or service provided will meet certain specified standards of performance and that the product or service will always be consistent with the documented standards.

JIT (Just In Time) This was initially a manufacturing approach where materials are ordered to arrive just when required in the process, no output or buffer stocks are held, and the finished product is delivered direct to the customer. Lean Sigma incorporates the principles of JIT and relates to the supply chain from supplier and supplier's supplier, through the process to the customer and the customer's customer.

Kaizen *Kaizen* is a Japanese word derived from a philosophy of gradual day by day betterment of life and spiritual enlightenment. This approach has been adopted in industry and means gradual and unending improvement in efficiency and/or customer satisfaction. The philosophy is doing little things better so as to achieve a long-term objective.

Kanban *Kanban* is the Japanese word for card. The basic *kanban* system is to use cards to trigger movements of materials between operations in production so that a customer order flows through the system. Computer systems eliminate the need for cards, but the principle is the same. As a job flows through the factory, completion of one stage of production triggers the next so that there is no idle time, or queues, between operations. Any one job can be tracked to determine the stage of production. A *kanban* is raised for each customer order. The *kanban* system enables production to be in batches of one.

KPIs (Key Performance Indicators) Measurements of performance, such as asset utilization, customer satisfaction, cycle time from order to delivery, inventory turnover, operations costs, productivity, and financial results (return on assets and return on investment).

Lean Sigma (see also see **JIT**) Lean was initially a manufacturing approach where materials are ordered to arrive just when required in the process, no output or buffer stocks are held, and the finished product is delivered direct to the customer. Lean Sigma incorporates the principles of Six Sigma, and is related to the supply chain from supplier and supplier's supplier, through the process to the customer and the customer's customer.

Mistake-proofing This refers to making each step of production mistake-free, and is also known as *Poka Yoke*. *Poka Yoke* was developed by Shingo (also see **SMED**), and has two main steps: (1) preventing the occurrence of a defect, and (2) detecting the defect. The system is applied at three points in a process:

1. In the event of an error, to prevent the start of a process
2. To prevent a non-conforming part from leaving a process
3. To prevent a non-conforming product from being passed to the next process.

MRP (II) (Manufacturing Resource Planning) Manufacturing resource planning is an integrated computer-based procedure for dealing with all of the planning and scheduling activities for manufacturing, and includes procedures for stock re-order, purchasing, inventory records, cost accounting, and plant maintenance.

Mudas *Muda* is the Japanese for waste or non-value-adding. The seven activities that are considered are:

1. Excess production
2. Waiting
3. Conveyance
4. Motion
5. Process
6. Inventory
7. Defects.

(For further detail, see Chapter 1.)

OEE (Overall Equipment Effectiveness) This is used to calculate the effective performance of an equipment and identify losses. In total productive maintenance (TPM) it is defined by the following formula:

OEE = Availability × Performance rate × Quality rate

Pareto Wilfredo Pareto was a nineteenth-century Italian economist who observed that 80 per cent of the wealth was held by 20 per cent of the population. The same phenomenon can often be found in quality problems. Juran (1988) refers to the vital few and the trivial many. The technique involves collecting data of defects, and identifying which occur the most and which result in the most cost or damage. Just because one defect occurs more often than others does not mean it is the costliest or that it should be corrected first.

PDCA (Plan–Do–Check–Act) The PDCA cycle was developed by Dr W.E. Deming, and refers to: Planning the change and setting standards; Doing (making the change happen); Checking that what is happening is what was intended (i.e. that standards are being met): and Acting – taking action to correct back to the standard.

Performance Charts or UCL/LCL Upper control and lower control limits are used to show variations from specification. Within the control limits, performance will be deemed to be acceptable. The aim should be to reduce the control limits over time, and thus control charts are used to monitor processes and the data gathered from the charts should be used to force never-ending improvements. These types of charts might also be known as Tolerance charts.

Poka Yoke This refers to making each step of production mistake-free, and is also known as mistake-proofing. *Poka Yoke* was developed by Shingo (also see **SMED**), and has two main steps: (1) preventing the occurrence of a defect, and (2) detecting the defect. The system is applied at three points in a process:

1. In the event of an error, to prevent the start of a process
2. To prevent a non-conforming part from leaving a process
3. To prevent a non-conforming product from being passed to the next process.

QFD (Quality Function Deployment) A systematic approach of determining customer needs and designing the product or service so that it meets the customer's needs first time and every time.

Qualitative Uses judgement and opinions to rate performance or quality. Qualitative assessment attempts to 'measure' intangibles such as taste, appearance, friendly service etc.

Quality Circles Quality circles are teams of staff who are volunteers. The team selects issues or areas to investigate for improvement. To work properly, teams have to be trained first in how to work as a team (group dynamics) and secondly in problem-solving techniques.

Quality Project Teams A top-down approach to solving a quality problem. Management determines a problem area and selects a team to solve the problem. The advantage over a Quality Circle is that this as a focused approach, but the disadvantage might be that members are conscripted rather than being volunteers.

Quantitative Means that which is tangible or can be measured – for example, the speedometer on a car measures and shows the speed.

RTY (Rolled Throughput Yield), *aka* **FPY (First Pass Yield)** This is the ratio of the number of completely defect-free units (without any kind of rework during the process) at the end of a process and the total number of units at the start of a process. The theoretical throughput rate is often regarded as the number of units at the start of the process. RTY/FPY is used as a key performance indicator to measure overall process effectiveness.

5 Ss These represent a set of Japanese words for excellent house keeping (*Sein*, sort; *Seiton*, set in place; *Seiso*, shine; *Seiketso*, standardize; and *Sitsuke*, sustain.

S & OP (Sales and Operations Planning) This is derived from MRP, and includes new product planning, demand planning, supply review to provide weekly and daily manufacturing schedules, and financial information (see also **MRP (II)**). S & OP is further explained in Chapter 6 (see Figure 6.10).

Scatter diagram Scatter diagrams are used to examine the relationship between two variables. Changes are made to each, and the results of changes are plotted on a graph to determine cause and effect.

Sigma The sign used for standard deviation from the arithmetic mean. If a normal distribution curve exists, one sigma represents one standard deviation either side of the mean and accounts for 68.27 per cent of the population. This is more fully explained in Chapter 3.

SIPOC (Supplier Input Process Output Customer) This represents a flow diagram and is used to define and examine an operation from the supplier to the customer.

Six Sigma Six Sigma is quality system that in effect aims for zero defects. Six Sigma in statistical terms means six deviations from the arithmetic mean, which equates to 99.99966 per cent of the total population, or 3.4 defects per million opportunities.

SMED (Single Minute Exchange of Dies) This was developed for the Japanese automobile industry by Shigeo Shingo in the 1980s, and involves the reduction of changeover of production by intensive work study to determine in-process and out-process activities and then systematic improve the planning, tooling, and operations of the changeover process. Shingo believed in looking for simple solutions rather relying on technology.

SoQ (Signature of Quality) A self-assessment process supported by a checklist covering customer focus, innovation, personnel and organizational leadership, use of technology, and environment and safety issues. It is useful in FIT SIGMA for establishing a company 'health' report.

SPC (Statistical Process Control) SPC uses statistical sampling to determine if the outputs of a stage or stages of a process are conforming to a standard. Upper and lower limits are set, and sampling is used to determine if the process is operating within the defined limits.

The Seven Wastes (see also *Muda*) *Muda* is the Japanese for waste or non-value-adding. The seven activities that are considered are:

1. Excess production
2. Waiting
3. Conveyance
4. Motion
5. Process
6. Inventory
7. Defects.

(See also Chapter 1.)

Tolerance charts or **UCL/LCL** Upper control and lower control limits are used to show variations from specification. Within the control limits performance will be deemed to be acceptable. The aim should be over time to reduce the control limits. Thus control charts are used to monitor processes and the data gathered from the charts should be used to force never ending improvements. These types of charts might also be known as Performance Charts.

TPM Total Productive Maintenance requires factory management to improve asset utilization by the systematic study and elimination of major obstacles – known as the 'six big losses' – to efficiency. The 'six big losses' in manufacturing are breakdown, set up and adjustment, minor stoppages, reduced speed, quality defects, and start up and shut down.

TQM Total Quality Management, is not a system, it is a philosophy embracing the total culture of an organization. TQM goes far beyond conformance to a standard, it requires a culture where every member of the organization believes that not a single day should go by without the organization in some way improving its efficiency and/or improving customer satisfaction.

UCL/LCL Upper control and lower control limits are used to show variations from specification. Within the control limits, performance will be deemed to be acceptable. The aim should be to reduce the control limits over time, and thus control charts are used to monitor processes and the data gathered from the charts should be used to force never-ending improvements. These types of charts might also be known as Performance charts, or Tolerance charts.

World class The term used to describe any organization that is making rapid and continuous improvement in performance and is considered to be using 'best practice' to achieve world-class standards.

Zero defects Philip Crosby made this term popular in the late 1970s. The approach is right thing, right time, right place, and every time. The assumption is that it is cheaper to do things right the first time.

Index

Bold type indicates discussion of the subject in detail

For Product Safety Concerns and Information please contact our EU representative GPSR@taylorandfrancis.com Taylor & Francis Verlag GmbH, Kaufingerstraße 24, 80331 München, Germany

T - #0087 - 230425 - C0 - 234/156/11 - PB - 9780750655613 - Gloss Lamination